Lecture Notes in Computer Science 7166

Commenced Publication in 1973
Founding and Former Series Editors:
Gerhard Goos, Juris Hartmanis, and Jan van Leeuwen

Editorial Board

Fernando A. Kuipers   Poul E. Heegaard (Eds.)

Self-Organizing Systems

6th IFIP TC 6 International Workshop, IWSOS 2012
Delft, The Netherlands, March 15-16, 2012
Proceedings

 Springer

Volume Editors

Fernando A. Kuipers
Delft University of Technology
Faculty of Electrical Engineering, Mathematics and Computer Science
P.O. Box 5031, 2600 GA Delft, The Netherlands
E-mail: f.a.kuipers@tudelft.nl

Poul E. Heegaard
Norwegian University of Science and Technology
Department of Telematics
O.S. Bragstads plass 2B, 7491 Trondheim, Norway
E-mail: poul.heegaard@item.ntnu.no

ISSN 0302-9743   e-ISSN 1611-3349
ISBN 978-3-642-28582-0   e-ISBN 978-3-642-28583-7
DOI 10.1007/978-3-642-28583-7
Springer Heidelberg Dordrecht London New York

Library of Congress Control Number: 2012932131

CR Subject Classification (1998): C.2, D.4.4, C.2.4, C.4, H.3, H.2.8, I.2.11

LNCS Sublibrary: SL 5 – Computer Communication Networks and Telecommunications

Typesetting: Camera-ready by author, data conversion by Scientific Publishing Services, Chennai, India

Printed on acid-free paper

Springer is part of Springer Science+Business Media (www.springer.com)

Preface

This book contains research articles from the IFIP International Workshop on Self-Organizing Systems (IWSOS), held in Delft, The Netherlands, in March 2012.

This was the sixth workshop in a series of multidisciplinary events dedicated to self-organization in networked systems with the main focus on communication and computer networks. The concept of self-organization is becoming increasingly popular in various branches of technology. A self-organizing system may be characterized by global, coordinated activity arising spontaneously from local interactions between the system's components. This activity is distributed over all components, without a central controller supervising or directing the behavior. Self-organization relates the behavior of the individual components (the microscopic level) to the resulting structure and functionality of the overall system (the macroscopic level). Simple interactions at the microscopic level may give rise to complex, adaptive, and robust behavior at the macroscopic level.

The necessity of self-organization in technological networks is caused by the growing scale, complexity, and dynamics of future networked systems. This is because traditional methods tend to be reductionistic, i.e., they neglect the effect of interactions between components. However, in complex networked systems, interactions cannot be ignored, since they are relevant for the future state of the system. In this sense, self-organization becomes a useful approach for dealing with the complexity inherent in networked systems. Since self-organization principles do not only apply to the Internet, but also to a variety of other complex networks, like transportation networks, telephony networks, smart electricity grids, smart cities, financial networks, social networks, and biological networks, IFIP IWSOS 2012 also invited articles on "Network Science" and "Complex Networks Theory" that focus on self-organizing systems.

The IFIP IWSOS 2012 committee received 28 submissions from 16 countries. Three papers were immediately rejected due to wrong scope or non-conformance with the template. Each of the remaining 25 submissions found to be eligible was in general reviewed by three members (papers with only two reviews had consistent reviews) of the Technical Program Committee. In total, 65 reviews were performed. Based on these reviews, ten papers were accepted. Five full papers were accepted out of the 20 full papers submitted, i.e., 25% acceptance rate. In addition, five short papers were accepted, which included three out of the eight short paper submissions, plus two full papers that were accepted as short papers. Two additional publications were invited for the workshop. The authors of the papers are from Italy, The Netherlands, Czech Republic, Germany, UK, South Korea, USA, Israel, Austria, and Norway.

The following key IWSOS topics were addressed:

- Design and analysis of self-organizing and self-managing systems
- Inspiring models of self-organization in nature and society
- Structure, characteristics and dynamics of self-organizing networks
- Techniques and tools for modeling self-organizing systems
- Robustness and adaptation in self-organizing systems
- Self-organization in complex networks like peer-to-peer, sensor, ad-hoc, vehicular and social networks
- Control of self-organizing systems
- Decentralized power management in the smart grid
- Self-organizing group and pattern formation
- Self-organizing mechanisms for task allocation, coordination and resource allocation
- Self-organizing information dissemination and content search
- Risks and limits of self-organization

The IWSOS series brings together leading international researchers to create a visionary forum for discussing the future of self-organization. It addresses theoretical aspects of self-organization as well as applications in communication and computer networks and complex networks.

IFIP IWSOS 2012 was technically sponsored by IFIP—the International Federation for Information Processing—and cosponsored by the European Network of Excellence Euro-NF.

The IFIP IWSOS 2012 workshop featured two keynote talks given by Shlomo Havlin, Bar-Ilan University, Israel, and Karl Aberer, EPFL, Switzerland. In addition, two panel sessions were organized: (1) Self-Organizing Smart Grids, and (2) Self-Organized Resilience, as well as a poster session with a student research competition.

January 2012   Fernando A. Kuipers
                                                          Poul E. Heegaard

Organization

General Chairs

Piet Van Mieghem   Delft University of Technology,
The Netherlands

David Hutchison   Lancaster University, UK

Program Chairs

Fernando Kuipers   Delft University of Technology,
The Netherlands

Poul Heegaard   NTNU-Trondheim, Norwegian University of
Science and Technology, Norway

Publicity Chairs

Maria Kihl   Lund University, Sweden

Karin Anna Hummel   University of Vienna, Austria

Local Organization

Christian Doerr   Delft University of Technology,
The Netherlands

Huijuan Wang   Delft University of Technology,
The Netherlands

Wendy Murtinu-van Schagen   Delft University of Technology,
The Netherlands

Steering Committee

Hermann de Meer   University of Passau, Germany

David Hutchison   Lancaster University, UK

Bernhard Plattner   ETH Zurich, Switzerland

James Sterbenz   University of Kansas, USA

Randy Katz   UC Berkeley, USA

Georg Carle   TU Munich, Germany (IFIP TC6
Representative)

Karin Anna Hummel   University of Vienna, Austria

Technical Program Committee

Ozalp Babaoglu	University of Bologna, Italy
Yuriy Brun	University of Washington, USA
Christian Doerr	Delft University of Technology, The Netherlands
Raissa D'Souza	University of Calfornia, Davis, USA
Falko Dressler	University of Innsbruck, Austria
Stefan Dulman	Delft University of Technology, The Netherlands
Schahram Dustdar	Vienna University of Technology, Austria
Wilfried Elmenreich	University of Klagenfurt, Austria
Alois Ferscha	University of Linz, Austria
Kurt Geihs	University of Kassel, Germany
Carlos Gershenson	Universidad Nacional Autónoma de México, Mexico
Salima Hassas	University of Lyon, France
Boudewijn Haverkort	University of Twente, The Netherlands
Bjarne Helvik	Norwegian University of Science and Technology, Norway
Tom Holvoet	Katholieke Universiteit Leuven, Belgium
Guy Leduc	University of Liège, Belgium
Hein Meling	University of Stavanger, Norway
Juval Portugali	Tel Aviv University, Israel
Christian Prehofer	Fraunhofer ESK, Germany
Andreas Riener	Johannes Kepler Universität Linz, Austria
Hiroki Sayama	Binghamton University, USA
Marcus Schöller	NEC Europe Ltd., Germany
Caterina Scoglio	Kansas State University, USA
Paul Smith	Lancaster University, UK
Ioannis Stavrakakis	National and Kapodistrian University of Athens, Greece
Maarten van Steen	VU University Amsterdam, The Netherlands
Bosiljka Tadic	Jozef Stefan Institute, Slovenia
Marc Timme	Max Planck Institute for Dynamics and Self-Organization, Germany
Vito Trianni	National Research Council - CNR, Italy
Huijuan Wang	Delft University of Technology, The Netherlands

Reviewers

Ozalp Babaoglu	University of Bologna, Italy
Yuriy Brun	University of Washington, USA
Christian Doerr	Delft University of Technology, The Netherlands

Falko Dressler	University of Innsbruck, Austria
Schahram Dustdar	Vienna University of Technology, Austria
Wilfried Elmenreich	University of Klagenfurt, Austria
Rinde van Lon	Katholieke Universiteit Leuven, Belgium
Kurt Geihs	University of Kassel, Germany
Carlos Gershenson	Universidad Nacional Autónoma de México, Mexico
Shaza Hanif	Katholieke Universiteit Leuven, Belgium
Boudewijn Haverkort	University of Twente, The Netherlands
Patrick Heinrich	Fraunhofer ESK, Germany
Bjarne Helvik	Norwegian University of Science and Technology, Norway
Tom Holvoet	Katholieke Universiteit Leuven, Belgium
Guy Leduc	University of Liège, Belgium
Hein Meling	University of Stavanger, Norway
Ebisa Negeri	Delft University of Technology, The Netherlands
Christian Prehofer	Fraunhofer ESK, Germany
Andreas Riener	Johannes Kepler Universität Linz, Austria
Hiroki Sayama	Binghamton University, USA
Marcus Schöller	NEC Europe Ltd., Germany
Paul Smith	Lancaster University, UK
Ioannis Stavrakakis	National and Kapodistrian University of Athens, Greece
Maarten van Steen	VU University Amsterdam, The Netherlands
Bosiljka Tadic	Jozef Stefan Institute, Slovenia
Vito Trianni	National Research Council - CNR, Italy
Stijn Vandael	Katholieke Universiteit Leuven, Belgium
Constantinos Vassilakis	Greek Research and Technology Network (GRNET), Greece
Huijuan Wang	Delft University of Technology, The Netherlands
Kashif Zia	University of Linz, Austria

Table of Contents

Distributed Storage Management Using Dynamic Pricing in a Self-Organized Energy Community

Ebisa Negeri and Nico Baken

Delft University of Technology, Delft, The Netherlands
{E.O.Negeri,N.H.G.Baken}@tudelft.nl

Abstract. We consider a future self-organized energy community that is composed of "prosumer" households that can autonomously generate, store, import and export power, and also selfishly strive to minimize their cost by adjusting their load profiles using the flexibly of their distributed storage. In such scenario, the aggregate load profile of the self-organized community is likely to be volatile due to the flexibility of the uncoordinated selfish households and the intermittence of the distributed generations. Previously, either centralized solutions or cooperation based decentralized solutions were proposed to manage the aggregate load, or the load of an individual selfish household was considered. We study the interplay between selfish households and community behavior by proposing *a novel dynamic pricing model* that provides an *optimal price vector* to the households to *flatten the overall community load profile*. Our dynamic pricing model intelligently adapts to the intermittence of the DGs and the closed-loop feedback that might result from price-responsiveness of the selfish households using its learning mechanism. Our dynamic pricing scheme has distinct import and export tariff components. Based on our dynamic pricing model, *we propose a polynomial-time distributed DS scheduling algorithm* that runs at each household to solve a cost minimization problem that complies with the selfish nature of the households. Our simulation results reveal that *our distributed algorithm achieves up to 72.5% reduction* in standard deviation of the overall net demand of the community compared to a distributed scheduling with two-level pricing scheme, and also gives comparable performance with a reference centralized scheduling algorithm.

1 Introduction

With the growing concerns for sustainable energy, an increasing number of diverse types of small-scale distributed generations (DGs), such as solar panels, are penetrating into neighborhoods. As a consequence, households are evolving from passive consumers towards active prosumers (producer-consumer) that can autonomously generate, store, export or import power. According to the EU parliament, all buildings built after 2019 will have to produce their own

F.A. Kuipers and P.E. Heegaard (Eds.): IWSOS 2012, LNCS 7166, pp. 1–12, 2012.

energy on site[1]. As households take power into their own hands, they tend to self-organize and form an "energy community" [1], for example at neighborhood level, to locally exchange power with each other and to influence the market with "the power of the collective." In such an autonomous community, the aggregate community load profile is likely to be highly volatile mainly because of two reasons. Firstly, the production of the DGs is highly intermittent. Secondly, the households could use the flexibility offered by their distributed storages (DSs), through charging and discharging, to selfishly adjust their load profile to minimize their cost, which could add up to a volatile aggregate profile such as large peaks that result if many households charge their DS at the same time. Indeed, DSs are becoming more attractive at the household level, especially along with intermittent DG, because they can increase demand-side flexibility and decrease the consumers' costs for electricity supply [2]. The volatility has various disadvantages. Firstly, a large number of high cost and carbon intensive "peak-plants" are required to provide power alongside the intermittent sources to compensate the variability. Moreover, the power peaks resulting from the variability may lead to infrastructure damages and expensive capacity upgrading. Thus, leveling the volatile overall community profile is crucial.

There are a few works in the literature on scheduling a system of multiple storage units ([3]-[4]). In these works, however, either the storage units are centrally scheduled without obeying the autonomy of the households, or households cooperatively play a game in a decentralized way with similar learning behavior to achieve a desired communal profile. Recently, dynamic pricing based demand-response strategy is gaining attention as a coordination mechanism. When the households are very flexible (for instance, by using their DSs), their responsiveness to the dynamic price might lead to a closed-loop feedback that could cause volatility, as demonstrated in [5]. Thus, in addition to reflecting the aggregate load profile, a dynamic pricing model needs intelligence to capture the flexibility of the households to respond to price signals.

In this paper, we consider a self-organized neighborhood energy community that is composed of selfish prosumer households, where power can be exchanged horizontally between the households within the community, and bidirectionally between the autonomous energy community and the rest of the grid. The energy community is controlled using a multi-agent based electronic market proposed in [6], where the households can trade with each other using local tariffs. A household-agent coordinates the resources at the household to optimize the household consumption, while a community-agent watches over the overall load profile of the energy community and coordinates the household-agents accordingly. The community is assumed to have smart grid capabilities [7], such as intelligent devices at each household that can exchange information across a communications platform.

The main contributions of this paper are:

---

[1] European Parliament, "All New Buildings to be Zero Energy from 2019," Committee on Industry, Research and Energy, Brussels 2009.

- We study coordination in a future energy community that is formed by a group of self-organized selfish prosumer households that can generate, store, import and export power, whereby the aggregate load profile is likely to be highly volatile due to the flexibility and selfish behavior of the prosumer households as well as the intermittence of DGs.
- We propose a dynamic pricing model that embodies a learning mechanism to iteratively adapt to the responsiveness of the households in the self-organized energy community and the intermittence of the DGs. Our pricing model has distinct import and export tariffs that handle the power import and export of the prosumers.
- We propose a polynomial-time distributed DS scheduling algorithm that runs at each household to solve a cost minimization problem of the selfish household, and at the same time flattening the overall community load profile using our dynamic pricing model.
- Our simulation results reveal that our distributed scheduling algorithm gains comparative performance with a reference centralized scheduling algorithm, and also achieves up to 72.5% reduction in standard deviation of the overall net demand of the community compared to a two-level pricing scheme.

The rest of the paper is structured as follows. Section 2 presents the mathematical model of the DS scheduling problem. In Section 3, the dynamic pricing model is presented, followed by our distributed algorithm in Section 4. The simulation results are presented in Section 5. Finally, our concluding remarks are presented in Section 6.

2   System Model

We present a formulation of the problem of finding an optimal schedule for DSs to flatten the demand profile of a self-organized neighborhood energy community. The distributed storage systems are scheduled offline over a scheduling time period τ (24 h) that is divided into time steps, i.e. $\tau = \{1, 2, ..., T\}$, that have equal duration of Δ time units.

2.1   Distributed Storage (DS)

Each DS is battery storage and is described by the parameters that are summarized below [9].

Φ: the maximum energy storage capacity (in kWh).
α: the maximum power capacity of the rectifier (in kW).
β: the maximum power capacity of the inverter (in kW).
ν: the time required for a full charge cycle (in h).
η^{st}: the cycle efficiency (%).
η^{rec}: the rectifier efficiency (%).
η^{inv}: the inverter efficiency (%).
δ: maximum depth of discharge allowed (%).

We neglect the self-dissipation of the battery, because it has a negligible effect for the length of our scheduling period. In addition to the above parameters, the variables X_j and Y_j are used to denote the amount of power that is charged into and discharged from, respectively, the DS in time step j. Assuming a symmetric cycle efficiency, η^{st}, we denote the storage charging efficiency and discharging efficiency by η^c and η^d, respectively, where $\eta^c = \eta^d = \sqrt{\eta^{st}}$. The state of charge (SOC) of the DS at time step j is denoted by Ψ_j (in kWh).

Generally, the SOC of a DS at the end of a time step k can be obtained from the SOC at the end of the previous time step j as shown in Eq. 1. The term added to Ψ_j accounts for the net rise in the storage level in time step k which is the difference between the stored energy due to charging of the storage and the discharged energy from the DS. The number of time units per time step (Δ) is used in Eq. 1 to convert power (kW) into energy (kWh). The maximum SOC of each DS is bounded by its maximum storage capacity (Eq. 2). The SOC cannot fall below the maximal discharging depth δ (Eq. 3). The rate of charging a DS is limited by the maximum charging rate ($\frac{\Phi}{\nu}$) and the maximum power capacity of the rectifier α (Eq. 4), whereas the discharging rate of the DS is limited by the maximum power capacity of the inverter β (Eq. 5).

$$\Psi_k = \Psi_j + \left(\Delta\eta^{rec}\eta^c X_k - \Delta\frac{Y_k}{\eta^{inv}\eta^d} \right) \tag{1}$$

$$\Psi_j \leq \Phi \, , \, \forall j \in \tau \tag{2}$$

$$\Psi_j \geq (1 - \delta) \times \Phi \, , \, \forall j \in \tau \tag{3}$$

$$X_j \leq \min(\alpha, \frac{\Phi}{\nu}) \, , \, \forall j \in \tau \tag{4}$$

$$Y_j \leq \beta \, , \, \forall j \in \tau \tag{5}$$

2.2   Demand and Production

The accumulated power demand of all the appliances of a given household in time step j is denoted by D_j, whereas we designate the total power production of all the distributed sources owned by the household in time step j by P_j. In this work, we assume that forecasts of D and P are available for the entire scheduling period. We define the *scheduled demand* of a household in time step j, denoted by R_j, as the difference between the demand and the self supply of the household in the time step (Eq. 6). The amount of power the household imports from or exports to the grid cannot exceed the power capacity of the cable connecting the household to the grid, denoted by ω, as shown in Eq. 7. The scheduled demand of the overall neighborhood R_j^o in time step j is given by the sum of the scheduled demands of all the households in the neighborhood as shown in Eq. 8. We represent the net demand of the community ignoring the DSs by Q_j^o, for each time step j (Eq. 9).

$$R_j = (D_j + X_j) - (Y_j + P_j), \forall j \in \tau \tag{6}$$

$$R_j \leq \omega, \forall j \in \tau \tag{7}$$

$$R^o_j = \sum_i R_j, \ \forall i \text{ households} \tag{8}$$

$$Q^o_j = \sum_i (D_j - P_j), \ \forall i \text{ households} \tag{9}$$

2.3   The Cost

Since a household can import power from the grid as well as feed the surplus power into the grid, we will have two different tariffs. We denote the import tariff and the export tariff in time step j by θ_j and λ_j, respectively. The electricity cost of a household in time step j is denoted by C_j, and is given in Eq. 10. Clearly, C_j is a positive cost when power is imported ($Q_j > 0$), and is a negative cost (income) when power is exported ($Q_j < 0$).

$$C_j = \begin{cases} \theta_j \times Q_j, & \text{if } Q_j \geq 0 \\ \lambda_j \times Q_j, & \text{if } Q_j < 0 \end{cases}, \ \forall j \in \tau \tag{10}$$

2.4   Optimization Problems

Given a self-organized neighborhood energy community of N prosumer households, each with forecasted values of demands and self productions and owning a DS with known capacities and efficiency parameters, the problem of optimally scheduling the N DSs over a scheduling period τ to achieve a flattened overall community load profile is addressed differently by the centralized and distributed scheduling strategies.

The centralized scheduling strategy gives the community-agent the authority to centrally schedule all the DSs of the households in the community. The community-agent first computes the average value of the net demand of the community (Eq. 11), and then schedules the DSs trying to minimize the deviation of the overall scheduled demand of the community from the average value. The corresponding quadratic optimization problem is formulated in Eq. 12, where $\mathbf{X}_{1,..,N}$, and $\mathbf{Y}_{1,..,N}$ represent the charging and discharging schedules, respectively, of the N DSs. Although this approach does not comply with the autonomy and selfish nature of the households, we use it as a reference.

$$mean = \frac{1}{T} \sum_{j=1}^{T} Q^o_j \tag{11}$$

$$\underset{\substack{X_1 \ldots X_N \\ Y_1 \ldots Y_N}}{\text{minimize}} \sum_{j \in \tau} (R^o_j - mean)^2 \tag{12}$$

subject to:

Equations 1-9

In the distributed scheduling technique each household-agent autonomously schedules its DS to selfishly minimize the cost of that household. The community-agent coordinates the household-agents by providing a novel dynamic pricing based incentive to achieve a flattened overall profile of the self-organized energy community. While the linear programming formulation of the household cost optimization problem is given in Eq. 13, the dynamic pricing model and the details of the distributed algorithm are deferred to Sections 3 and 4, respectively.

$$\text{minimize}_{X,Y} \sum_{j \in \tau} C_j \qquad (13)$$

subject to: Equations 1-7

3   The Dynamic Pricing Model

Our dynamic pricing model is based on a scenario where the autonomous households trade electricity with each other on a local electronic energy market [6] in the community whereby the autonomy of the energy community allows the community-agent to propose local price-vectors to achieve desirable communal goals. In our dynamic pricing model, the community-agent proposes a price vector, then each household-agent selfishly schedules the storage unit of that household by solving Eq. 13 and replies with the corresponding scheduled demand of the household. Based on the response of all the households, the community-agent adjusts the price vector using an intelligent learning mechanism. This is repeated iteratively to obtain an optimal price vector that yields a flattened overall community profile.

In the proposed pricing model, the cost of a unit power varies from one time step to another. The optimal tariff for each time step is determined iteratively. At the $(k + 1)^{th}$ iteration cycle, the tariff for the jth time step is obtained by adjusting the corresponding tariff in the previous iteration cycle k using an incremental factor $(\gamma \times \xi_{k,j})$:

$$\theta_{k+1,j} = \theta_{k,j}(1 \ + \ (\gamma \times \xi_{k,j})) \qquad (14)$$

The incremental factor is composed of two terms: the *learning factor* (γ) and the *deviation factor* $(\xi_{k,j})$. The learning factor is a constant $(0 < \gamma \leq 1)$ that determines to which extent the deviation factor overrides the old price vector. Empirical optimal values of γ are provided in Section 5. The deviation factor captures the variation of R^o from its average value. Let $R^o_{k,j}$ be the overall scheduled demand of the neighborhood in the jth time step for the kth iteration cycle. Let $mean^*$ be the average of the overall scheduled demand (Eq. 15), then the deviation factor, $\xi_{k,j}$, is given in Eq. 16, where $sign$ determines whether the incremental value should be positive or negative. If $Q^o_{k,j} > mean^*$, then we add a positive incremental value to the tariff to reduce consumption in this time step: $sign = 1$. If $Q^o_{k,j} < mean^*$, then we subtract an incremental value from the tariff to increase consumption in this time step: $sign = -1$.

$$mean^* = \frac{1}{T} \sum_{i=1}^{T} R_{k,j}^o \tag{15}$$

$$\xi_{k,j} = sign \times \frac{(R_{k,j}^o - mean^*)^2}{\sum_{i=1}^{T}(R_{k,i}^o - mean^*)^2} \tag{16}$$

$\xi_{k,j}$ reflects the effect of the offset of the scheduled demand from the mean value on the price vector. The pricing model increases the tariff on the time steps where the overall scheduled demand of the neighborhood (R^o) is above the average, and reduces the tariff when R^o is below the average, thereby providing incentives to the selfish households to flatten the overall scheduled neighborhood demand. Since the price vector is improved iteratively depending on the reaction of the households, the intelligent learning employed by our dynamic pricing model effectively handles the price-responsiveness of the households, thereby minimizing the closed-loop feedback effect. In effect, the aggregate load profile of the self-organized energy community could be managed.

Following the suggestion in [8], the power feed-in tariffs are obtained by subtracting the transport part of the import tariffs (Eq. 17), where θ^{tr} is the transport part of the import tariff θ_j. The pricing model can be modified to achieve objectives other than flattening the overall profile by simply substituting the $mean^*$ in Eq. 16 by the desired target value for each time step.

$$\lambda_j = \theta_j - \theta^{tr}, \forall j \in \tau \tag{17}$$

4   The Distributed Algorithm

Our design of the distributed algorithm reflects the autonomy and selfishness of the households, while simultaneously striving to flatten the aggregate load profile of the self-organized energy community using our novel dynamic pricing model. The block diagram describing the execution of our distributed scheduling algorithms is shown in Fig. 1. The algorithm embodies a fixed number of iterative cycles that search for an optimal price vector that yields the best level of flatness for the overall scheduled net demand of the energy community (R^o). We describe the flatness of R^o in terms of its standard deviation:

$$\sigma = \sqrt{\frac{1}{T} \sum_{j=1}^{T}(R_j^o - mean^*)^2}$$

The algorithm is initiated at the community-agent. At each iteration step, the community-agent sends the recent tariff vectors (θ, λ) to each household-agent. Upon receiving the tariff vectors, each household-agent solves its selfish cost optimization problem (Eq. 13) with the up-to-date tariffs, and sends its resulting scheduled demand to the community-agent. When the community-agent receives the scheduled demands of all the households in the current iteration cycle, it computes the tariff vectors for the next iteration cycle using Eq. 14, and the iteration continues. Once the desired fixed number of iterations is performed, the community-agent picks the best tariff vectors that yielded the smallest value of σ, and the corresponding schedules of the DSs are adopted.

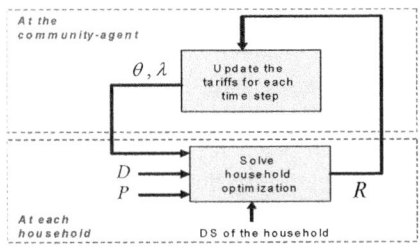

Fig. 1. A block diagram for the distributed algorithm

5   Simulation

In our simulation, storage systems installed at all the households have the same parameter specifications. Lead-acid battery storage is used as a reference, since it is regarded as the most economical choice [9]. The parameter specifications for lead-acid suggested in [9] are listed in Table 1. The data for the demand and electricity production of a household are acquired from Alliander[2]. The distributed sources of a household are a PV panel and a micro-CHP. The data were obtained by field measurements that were taken every 15 minutes for a duration of 24 hours using smart meters installed at the households. The micro-CHP generation is constant (1 kW) over each period of operation, and the data specifies the start time and end time of each period of its operation. To make alternative profiles for each household, we randomized the values according to a normal distribution. For demand and PV panel production, the randomized profiles were generated with mean value equal to the measured data and standard deviation equal to 15% of the mean, for each time step. For the micro-CHP production, different variants of the start time and end time of the production periods are generated using the normal distribution function with mean values equal to the corresponding measured values and standard deviation of one time step (15 minutes). A sample of the daily profile of a household in Spring season is shown in Fig. 2.

The transport part of the electricity import tariff is $\theta^{tr} = 0.04/kWh$. The data is obtained from ECN[3]. Thus, for each time step j, the feed-in tariff is obtained as $\lambda_j = \theta_j - 0.04/kWh$.

---

[2] Alliander is the largest electricity distribution network operator company in the Netherlands owning 40% of the distribution networks.

[3] Energy Research Centre of The Netherlands, http://www.energie.nl/.

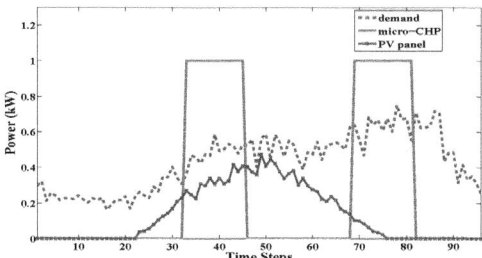

Fig. 2. Demand and production profiles of a household

Table 1. Parameter Values of the Storage System [1]

Φ	ν	δ	η^{st}	α	β	η^{rec}	η^{inv}
kWh	h	(%)	(%)	kW	kW	(%)	(%)
5	5	80	85	1	0.5	95	98

The proposed algorithm is simulated with different demand and generation data sets representing each of the four seasons of the year. In addition to the centralized scheduling and the dynamic pricing based distributed scheduling, we have a simulation scenario in which households autonomously schedule their DSs to minimize their cost responding to fixed two level tariffs (day and night tariffs).

The relative performance of the scheduling algorithms was similar under the data sets representing the four seasons, thus we present only the results obtained using the data sets representing the Spring season due to limited space. Fig. 3 shows the overall scheduled demand of the community obtained using the centralized and the distributed scheduling techniques. The large negative net demands of the neighborhood in the time steps 32-45 and 67-80 result from the accumulated production of the micro-CHPs and the PV panels that exceed the total demand. Apparently, both the centralized and distributed scheduling techniques have achieved significant improvement (75.6% and 75.0% reductions in σ, respectively) in the overall scheduled demand profile compared to the net demand of the community. It is interesting to see that the distributed technique, that maintains the autonomy and selfishness of the households, has achieved a comparable performance with the centralized technique that makes a simplifying assumption that all the DSs are centrally managed. The achievement of the distributed technique stems from the intelligence of its dynamic pricing model based incentive.

The comparison of the performance of our dynamic pricing based distributed scheduling technique to that of the two level pricing scheme shown in Fig. 4 reveals that our distributed algorithm achieves a profile with significant improvement in flatness compared to the two level pricing. Apparently, the scheduled demand using two price levels did not show significant improvement on the shape

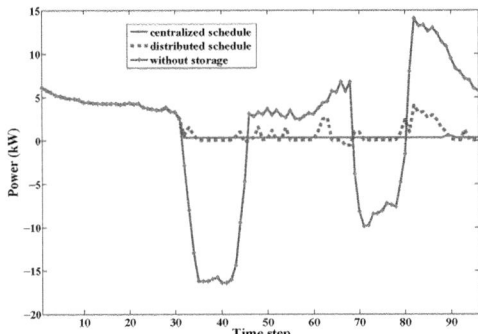

Fig. 3. Comparison of the centralized and distributed algorithms using the overall scheduled demand for N = 20

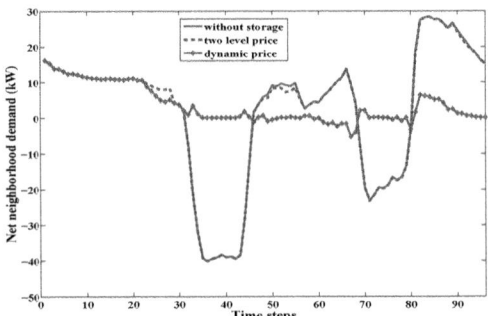

Fig. 4. Comparison of the dynamic pricing and two-level pricing schemes using the overall scheduled demand for N = 50

of the overall net demand, implying that the pricing scheme does not incentivise the households enough. Our simulation results based on the data representing the four seasons of a year reveal that our dynamic pricing based algorithm can achieve up to 72.5% improvement compared to the two-level pricing scheme in terms of the standard deviation of the overall scheduled demand.

A comparison of the performance of the distributed algorithm under different values of the learning factor (γ) is shown in Fig. 5. The figure shows that the standard deviation of the overall scheduled demand decreases slowly for smaller values of γ, whereas, it drops rapidly for larger values of γ. On the other hand, the standard deviation curve is more stable for smaller values of γ. Furthermore, no significant improvement in σ is observed after a fixed number of iterations. Hence, the algorithm is expected to perform well with few iterations using the larger values of γ ($0.2 \leq \gamma \leq 0.5$).

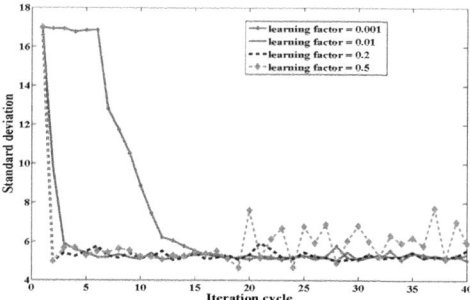

Fig. 5. The effect of the learning factor

Performance deviations (% value) due to parameter value variations

Parameter	Ref. value	Reference value variation					
		-20%	-10%	-5%	+5%	+10%	+20%
Cycle eff. η^{sf}	82	-4.6	-2.4	-1.3	1.4	5.2	6.8
Rectifier eff. η^{rec}	95	-7.9	-4.9	-0.4	0.4	n/a	n/a
Inverter eff. η^{inv}	98	-9.0	-1.2	-0.04	n/a	n/a	n/a
		-50%	-25%	-10%	+10%	+25%	+50%
Max. charg. rate, $\min(\alpha, \frac{\Phi}{\upsilon})$	1	-9.3	-0.03	-0.01	0.01	0.01	0.07
Max. disch. rate, β	0.5	-24.5	-7.3	-1.7	1.7	2.6	2.9
Community size, N	50	1	10	20	30	40	50
		-5.41	-0.97	-0.31	-0.12	-0.1	0
Storage capacity, Φ	5	1	2	3	7	9	11
		137	41	0	0	0	0

Fig. 6. Algorithm performance in response to perturbing parameters

The reference parameter values in Table 1 are perturbed to assess their impact on the performance of our distributed algorithm. The results in Fig. 6 reveal that varying the efficiency parameters, the maximum charging and discharging rates of the DS results in low variations (mostly between -10% and +10%) in the performance of the algorithm, where increasing the values of the parameters increases flexibility of the households to respond to the dynamic price, thereby slightly increasing performance. Varying the community size also resulted in small performance deviation which results form the randomization of the profiles of the households. On the other hand, with very low storage capacity, the households have limited flexibility to store and release power in response to the dynamic price, thereby leading to larger values of σ.

6   Conclusions

The load profile of a self-organized energy community composed of selfish prosumers that can autonomously generate, store, import, and export power, is likely to be highly volatile due to the intermittence of the distributed generations and the autonomy and selfish nature of the prosumers. We have introduced a novel dynamic pricing model that can help to flatten the load profile in such scenario by intelligently adapting to the price-responsiveness of the selfish prosumers using its learning mechanism. We have proposed a distributed scheduling algorithm of distributed power storages that uses the dynamic pricing model to flatten the community load profile while each household selfishly tries to minimize its cost by solving a simple polynomial-time optimization problem formulated via linear programming. Simulation results reveal that our distributed algorithm has comparable performance with a reference centralized scheduling algorithm, and it can achieve up to 72.5% improvement in standard deviation of the load profile of the community compared to a reference two-level pricing model.

Our distributed algorithm and dynamic pricing model can also be extended to schedule flexible electrical units such as electric vehicles, heat-pumps, etc., and could be easily modified for objectives other than flattening the overall demand.

References

1. Melo, H., Heinrich, C.: Energy Balance in a Renewable Energy Community. IEEE EEEIC (May 2011)
2. Mulder, G., Ridder, F., Six, D.: Electricity Storage for Grid-connected Household Dwellings with PV Panels. Solar Energy 84, 1284–1293 (2010)
3. Cau, T., Kaye, R.: Multiple Distributed Energy Storage Scheduling Using Constructive Evolutionary Programming. In: Proc. of IEEE Power Engineering Society International Conference, vol. 22, pp. 402–407 (August 2002)
4. Vytelingum, P., Voice, T., Ramchurn, S., Rogers, A., Jennings, N.: Agent-based Micro-Storage Management for the Smart Grid. In: Proc. of 9th Int. Conf. on Autonomous Agents and Mutiagent Systems (AAMAS 2010) (May 2010)
5. Roozbehani, M., Dahleh, M., Mitter, S.: On the Stability of Wholesale Electricity Markets Under Real-Time Pricing. In: IEEE Conference on Decision and Control (CDC), pp. 1911–1918 (December 2010)
6. Kok, J., Warmer, C., Kamphuis, I.: PowerMatcher: Multi-Agent Control in Electicity Infrastructure. In: AAMAS 2005 (July 2005)
7. Ipakchi, A., Albuyeh, F.: Grid of The Future. IEEE Power and Energy Magazine, 52–62 (March 2009)
8. Houwing, M., Negenborn, R., Schutter, B.: Demand Response with Micro-CHP systems. Proceedings of the IEEE 99(1), 200–212 (2011)
9. Ahlert, K., Dinther, C.: Sensitivity Analysis of the Economic Benefits from Electricity Storage at the End Consumer Level. IEEE Transactions on PowerTech, 1–8 (October 2009)

Ant-Based Systems for Wireless Networks: Retrospect and Prospects

Laurent Paquereau[1] and Bjarne E. Helvik[2]

[1] Department of Telematics,
Norwegian University of Science and Technology, Trondheim, Norway
laurent@item.ntnu.no
[2] Centre for Quantifiable Quality of Service in Communication Systems[*],
Norwegian University of Science and Technology, Trondheim, Norway
bjarne@q2s.ntnu.no

Abstract. Since they were first introduced as powerful stochastic optimization systems, ant-based systems have been applied to a wide range of problems. One of the most successful applications has been routing in dynamic wired telecommunication networks. Following this success, similar approaches have been applied to routing in multi-hop wireless networks. The objective has been to achieve self-organizing path management in these networks. This paper looks back on 10 years of research and presents reflections on the challenges, the evolution, the contributions and the future perspectives in this field.

1 Introduction

Ant-based systems are systems inspired by the foraging behavior of ants in nature or, more generally, systems that can be described as such. Ant-based systems are swarm intelligence systems. Each individual ant is not capable of finding the shortest path between its nest and a food source. It is by their collective behavior that ants in a colony are able to converge to the shortest path; a key enabler to this emergent behavior being the indirect communication between ants through the modification of their environment (stigmergy).

Ant-based systems were first developed as powerful stochastic optimization systems and shown to be able to achieve state-of-the-art results on complex combinatorial problems. Later, ant-based techniques were combined with the idea of using mobile agents for path management in next-generation telecommunication networks. The motivation was to apply stochastic optimization in the network and develop decentralized, adaptive, robust and self-organizing solutions to cope with the growing complexity of networks. On the basis of the promising results obtained for wired networks in the second half of the 1990's, ant-based systems for routing in multi-hop wireless networks (MHWNs) started to be developed, the first of them being ARA [16] proposed in 2002.

[*] "Centre for Quantifiable Quality of Service in Communication Systems, Centre of Excellence" appointed by The Research Council of Norway, funded by the Research Council, NTNU and UNINETT. http://www.q2s.ntnu.no

F.A. Kuipers and P.E. Heegaard (Eds.): IWSOS 2012, LNCS 7166, pp. 13–23, 2012.
© IFIP International Federation for Information Processing 2012

In the context, ants are control packets that are used to repeatedly sample paths between source and destination pairs. At each node, the pheromone trails reflect the knowledge acquired by the colony, and the pheromone trail value associated with a given neighbour indicates the relative goodness of this neighbour to reach a specific destination.

The objective with this paper is neither to present yet another extensive survey nor a detailed description of the inner working of ant-based systems for MHWNs. For that, the interested reader is referred to [15,18]. Rather, it is to look back on 10 years of research on applying ant-based techniques to routing in MHWNs and present some reflections on the challenges, evolutions, contributions, trends and perspectives in the field, touching upon both technical and non-technical aspects.

The rest of this paper is organized as follows. First, Section 2 addresses the application of ant-based techniques to mobile ad-hoc networks (MANETs), which is the most common type of MHWNs ant-based techniques have been applied to. Next, Section 3 discusses the evolution and current status of the research in the field and the perspectives for future developments and applications. Finally, concluding remarks are given in Section 4.

2   Applying Ant-Based Techniques to MANETs

This section addresses the application of ant-based techniques to MANETs. The focus is on unicast routing protocols as this represents the major part of the contributions, but there have also been proposals for using ant-based techniques for establishing and maintaining multicast trees in the context of MANETs, e.g. MANSI [23].

This section elaborates first on the premises for developing ant-based systems for routing in MANETs and the challenges related to the application of ant-based techniques in this context. The main solutions found in the literature are then presented. Finally, all these elements are summarized and discussed.

2.1   Premises

MANETs are autonomous MHWNs in which nodes communicate without the support of any pre-existing infrastructure and with no centralized control. Nodes may operate both as routers and hosts. They are all potentially mobile and battery-powered, and they may have limited processing and memory capacity. MANETs typically exhibit a high degree of dynamicity as nodes may move and leave or join the network at any time, making control, and in particular routing, a challenging task. Moreover, in addition to being able to deal with the changes in the environment, any control system designed for MANETs has to cope with the limited bandwith and the inherent unreliability of wireless communications, making this task even more challenging.

The main reasons that are put forward in the literature to support the application of ant-based techniques to routing in MANETs are: (i) the good match between the desirable properties of a path management systems for MANETs and the characteristics of ant-based systems in terms of adaptivity and robustness, and, as mentioned in the introduction, (ii) the promising results obtained applying ant-based techniques to routing in wired networks, in particular with AntNet [10]. Adaptivity is the ability of the system to adapt to changes and prevent disruptive events. This feature is empowered in ant-based systems by the repeated stochastic sampling of end-to-end paths and the resulting update of pheromone trail values at each node. It ensures the availability of timely routing information and enables the discovery and maintenance of multiple paths between source and destination pairs. Robustness is the ability of the system to maintain persitent operation under the occurrence of perturbations. Typical perturbations in MANETs include the loss of packets and the loss of routing information stored at nodes. Ant-based systems are robust in that they are not sensitive to the loss of individual ants and that routing information (pheromone trail values) is distributed throughout the network.

2.2 Challenges

The main challenge when applying ant-based techniques to routing in MANETs is efficiency. This is a two-sided challenge. Efficiency can be expressed in terms of routing overhead and in terms of convergence time. Others challenges include the proactive nature of ant-based systems, spreading the load, demands in terms of memory and processing and energy consumption.

Routing Overhead. The routing overhead for ant-based systems is related to the repeated probing of paths realized by ants. This implies forwarding control packets back and forth between source and destination pairs which does consume bandwidth. As the wireless medium is shared and the available bandwidth is limited, controlling the overhead is therefore crucial for the performance of the system (data delivery ratio, delay, delay jitter). However, there is a trade-off between performance on the one side, and adaptivity and robustness on the other side. Ant-based systems are intrinsically robust to the loss of individual ants because they generate many of them. Reducing the number of ants directly affects the robustness of the system. In addition, the ability of the system to discover new paths and adapt to topological changes is conditioned by the fact that ants sample paths and update pheromone tables accordingly. Reducing the number of ants also directly affects the adaptivity of the system.

Convergence Time. Ant-based systems apply stochastic optimization and several iterations are required before such a system converges to a solution. Iterations corresponds to updates made by ants. Hence, the time it takes for a system to converge is related to the rate at which ants are generated. To be able to adapt, the time scale at which changes take place in the network should be much larger than that at which pheromone trail values are updated. Since MANETs are assumed to be highly dynamic environments and the number of ants required to keep up with the changes may be prohibitive with respect to the routing overhead incurred.

Proactive Nature. Ant-based systems rely on the proactive sampling of paths and comsume resources forwarding ants and maintaining pheromone tables at each node. In a wired networks, this makes sense as the topology is seldom changing, and, although nodes or links might failed, it is expected that they are restored with the same characteristics so that previously gathered data are still relevant and can actually be used to re-converge faster. In the context of MANETs, spending resources for proactive planning does generally not pay off as the conditions, in terms of topology, wireless link characteristics and traffic demands, are constantly changing. In addition, keeping historical data may be harmful for performance. When a node loses connectivity with one of its neighbor, although connectivity might be re-established at a later point in time with the same node, there is no guarantee that the quality of the path through this neighbor will be similar. Previously collected data may no longer be relevant and may lead to reduced performance if used.

Load Spreading. The promising results obtained with AntNet in wired networks are in part due to its ability to achieve near-optimal load distribution by spreading the load along multiple path proportionally to their estimated quality. Achieving similar results in the context of MANETs is challenging for several reasons: (i) efficiently spreading the load in a highly dynamic environment is difficult, (ii) the behavior of the nodes and the traffic patterns in MANETs are unpredictable hence no assumptions that would help in distributing the load can be made, and (iii) distributing the load across multiple paths in MHWNs may lead to degraded performance because of interferences if the paths are not radio-disjoint [12,20].

Memory, Processing and Energy Constraints. Ant-based systems rely on the repeated and stochastic sampling of paths by ants and the maintenance of pheromone tables locally at each node. At each node, the pheromone table potentially contains an entry for each destination in the network through each of the neighbouring nodes. This implies repeated computations on a per-packet basis and imposes requirements in terms of processing and storage at each node. In addition, performing these operations and regularly forwarding ants are energy consuming tasks.

2.3   Solutions

Ant-based systems developed for MANETs are mostly derived from AntNet or directly apply the Ant Colony Optimization (ACO) meta-heuristic [11]. Systems vary among others in terms of formulas used for pheromone updates, cost functions (hop-count, delay, or some measure of link quality including signal-to-noise ratio, congestion or packet loss ratio, or any combination of those) and mechanisms used to improve the performance. Studies such as [3,9] have shown that directly applying ant-based systems developed for wired networks such as AntNet results in lower performance compared to conventional MANETs protocols, and in particular to AODV [21], which is used as the reference protocols in most of the studies. In the following, the focus is on the mechanisms introduced specifically to address the challenges listed in Section 2.2.

Broadcasting Forward Ants. Most of the systems broadcast forward ants to explore multiple routes concurrently and, hence, reduce the convergence time. However, this approach has also several downsides: (i) paths are sampled regardless of their estimated quality, (ii) it results in a large overhead as ants get duplicated at each node, (iii) it results in the establishment of paths with low quality, and (iv) it leads to the discovery and reinforcement of a set of braided (interfering) paths that cannot therefore not be used to improve the performance of system by distributing the load. The resulting process is similar to the route discovery process of AODV, and some systems also introduce mechanisms used by AODV to limit the overhead such as expanding ring-search. To limit the overhead other systems such as AntHocNet [13] resolve to broadcast only when no routing information is available at a node and otherwise apply a random proportional rule and unicast the ant only to one of the neighboring nodes.

Adopting a Greedy Deterministic Forwarding Policy. Most of the systems abandon the stochastic forwarding policy used in AntNet for distributing the load accross multiple path and instead adopt a greedy deterministic forwarding policy, i.e. each node forwards data packets to its neighbor with the highest pheromone concentration for the given destination. The main reasons, as mentioned above, are that it is difficult to distribute the load in a dynamic environment and that distributing the load might lead to degraded performance because of interferences. Authors in [12] also show that the best performance with AntHocNet is achieved when a greedy deterministic policy is applied to the forwarding of ants. The reason is that the number of ants that can be generated is often too low for the system to be able to cope with the constantly changing environment, and the system cannot effectively explore multiple paths.

Relying on Data Packets for Updating Pheromone Values. In order to reduce the overhead, some of the proposed systems rely partly or solely on data packets to update pheromones, e.g. ARA and SARA [6]. The major drawback of this approach is that it affects directly the adaptivity and robustness of the system as it becomes dependent on data traffic to maintain timely routing information. Using data packets for exploration also results in decreased performance in terms for instance of increased delay and jitter.

Combining Reactive and Proactive Path Sampling. To reduce the overhead, most of the proposed systems adopt a reactive strategy, meaning that a node starts generating ants only when it is itself source of data traffic. Many systems keep, however, a proactive component and are therefore classified as hybrid, e.g. AntHocNet. The information collected by proactive ants may then used for bootstraping the reactive route discovery process and/or for improving the path and adapting to changes while data sessions are active.

Introducing Additional Control Packets for Link Error Notification and Handling. To improve the adaptivity of the system, many systems introduce additional dedicated control packets for explicit link failure notifications, local recovery and/or to deal with lost link failure notifications, e.g. ARA and AntHocNet. One disadvantage of these mechanisms is that they create an extra overhead upon link failure. Another aspect is that local recovery techniques are opposite with the objective of converging towards a global optimum by constructing complete solutions, which is one of the advantages of ant-based systems as a stochastic optimization method.

Introducing Pheromone Diffusion. This mechanism is introduced in AntHocNet. The idea is to let nodes share routing information with their 1-hop neighborhood via asynchronous beacon messages and derive virtual pheromones that are used to guide the exploration of ants.

2.4   Discussion

To deal with the challenges listed in Section 2.2 and bear the comparison with state-of-the-art reactive MANET routing protocols, in particular AODV, most of the proposed systems include mechanisms borrowed from those same protocols. This introduces a deviation from key elements of the optimization method (proactive sampling of end-to-end solutions, stochastic forwarding policy) and from what made its successful application to dynamic routing problem in wired networks (load-spreading). As a result, most of the systems are intermediate between conventional reactive MANET routing protocols and ant-based systems developed for path management in wired networks, such as AntNet. Some of them are much closer to AODV although the terminology used is that of ant-based systems.

We argue that ant-based techniques have been applied to MANETs on wrong premises and question the suitability of this approach for routing in MANETs in the first place. The adaptability and robustness properties of an ant-based systems stem from the repeated sampling of path by ants, but are conditioned by the fact that the intensity of changes the system has to adapt to is sufficiently low compared to the rate at which ants sample paths and update pheromone trails. In addition, there is a trade-off between how adaptive and robust an ant-based system can be and the routing overhead it generates. In typically settings for MANETs, changes occur too fast for the overhead not to be prohibitive.

3   Evolution, Status and Perspectives

In this section, we first present the evolution and the current status of the research on ant-based techniques for wireless networks. We then discuss when this approach is viable and when it is no, the question of the future of this approach and sketch possible future directions.

3.1   Evolution

The development of the first generation of ant-based systems for MHWNs, targeting MANETs, started at the beginning of the 2000's; the first ant-based system for MANETs reported in the literature being ARA [16]. The explanation of why it started at that time is that: (i) routing in MANETs was a hot topic in the second half of the 1990's and early in 2000, and that, concurrently, (ii) promising results achieved by ant-based systems for wired networks were published by artificial intelligence researchers and attracted the attention of the MANET research community. Reversely, the development of ant-based systems for MANETs by networking researchers later attracted the attention of the artificial intelligence community (AntHocNet was first reported in 2004). Another reason is (iii) the observation made by De Couto et al. in 2002 [8] that minimizing the number of hops to decide on the best route between nodes in a MHWN is not necessarily synonym of maximizing the performance as longer links are generally more unstable than shorther ones. The fact that it is possible to integrate the link/path quality in the computation of the pheromone trail values is listed as an argument for applying ant-based techniques to MANETs in [16].

Starting from 2004, and following the trend observed in MHWN research, a second generation of ant-based systems, targeting specialized MHWNs, started to be developed. The reason for this specialization is that the definition of MANETs is very generic and encompasses a wide range of scenarios and applications, which cannot all be best solved by a single protocol design. In particular, ant-based systems were developed for Wireless Sensor Networks (WSNs) [2] and backbone Wireless Mesh Networks (WMNs) [1]. The characteristics of these specialized MHWNs and of generic MANETs are summarized in Table 1.

Table 1. Characteristics of MHWNs

MHWNs	Limited power, memory, processing	Mobility	Predictable traffic patterns	High rate of changes in the network	Limited bandwidth	Inter-ferences
MANET	X	X		X	X	X
WSN	X		X		X	X
WMN			X		X	X

In particular, a relatively large number of ant-based systems targeting WSNs have been designed, see for instance [22] for a survey. WSNs have characteristics that make proactive approaches and load spreading a priori more appropriate than in generic MANETs: (i) nodes are static and, (ii) certain traffic patterns are expected between sensor nodes and sink nodes making load distribution both easier and more relevant. Compared to generic MANETs, restrictions in terms of memory and processing capacity as well as energy constraints are however more compelling. Hence, one can also question the intrinsic suitability of ant-based approaches in this context.

Similarly to WSNs, nodes in WMNs are static and the traffic is expected to be concentrated on certain paths (to and from gateway nodes providing connection to the Internet). On the other hand, contrary to WSNs, nodes in WMNs are grid-powered wireless routers with high processing and memory capacity. Therefore, although the number of contributions addressing specifically backbone WMNs is still limited, they seem to be the most actual type of MHWNs for the application of ant-based techniques. An example of such a system is for instance AMIRA [4].

Finally, it should be noted that the research in the field has mainly been be driven by method-oriented rather than problem-oriented researchers. As for wired networks with AntNet, the most significant work in the field to date, AntHocNet, has been contributed by authors with a background in artifical intelligence. Applying ant-based techniques to MHWNs was, in a way, the next step after the promising results achieved with AntNet.

3.2   Status

As we assess the state of art, the research in ant-based techniques for routing in MANETs is about to cease, because: (i) as concluded in Section 2, ant-based systems are intrinsically not suitable for MANETs, and (ii) although several of the proposed systems and in particular AntHocNet are shown to outperform AODV, the gain in performance is not worth the added complexity. This latter assertion is supported by the work of AntHocNet authors who first observed in [12] that stochastically sampling paths results in lower performance compared to deterministically resampling the best paths, and who later proposed a new approach for integrating proactive and reactive routing in MANETs that built up on observations they made with AntHocNet, but with no references to ant-based techniques [14].

The application of ant-based techniques to specialized MHWNs is, on the other hand, still an open research topic. In spite of the fact that one may question its suitability for WSNs, many such systems have been proposed. It is, however, hard to assess the performance these proposals, since comparisons with state-of-the-art algorithms specifically designed for WSNs are seldom provided. Furthermore, both real-world implementations and testing, as well as mathematical modeling, is also listed as missing in [22].

In the context of WMNs, few ant-based systems have been proposed so far. Pursuing these further, a first necessary step is a comparison with conventional state-of-the-art proactive routing protocols such as OLSR [17]. In the context of MANETs, ant-based systems have most of the time only been compared with AODV. To the best of our knowledge, no comparisons with traditional proactive MANET routing protocols have been reported.

3.3   Perspectives

Have ant-based techniques in MHWNs a future? If we expect it to be a simple, efficient and generally applicable technique, the answer is no. The technological rationale for this is found in Section 2. The method has hereto not been able to overcome the significant challenges put by the highly dynamical, wireless domain. On the other hand, we argue that the path management problem in WMNs is closer to that addressed in wired network than that addressed in MANETs, and that thus ant-based systems may have a future in this context. Moreover, there are specific aspects in the path management problem in WMNs for which ant-based systems are relevant and attractive candidates. For instance, mesh routers are usually assumed to be equiped with multiple wireless interfaces and the channel assignment problem presents some similarities with the graph coloring problem for which good results have been reported using ant-based systems in a static context [7].

Although the path management problems in WMNs is closer to those of wired network than those addressed in MANETs, the challenges related to wireless communication - limited bandwidth and interferences - remain. Hence, applying systems developed for wired networks directly to WMNs leads to poor performance. One observation is that to rectify the problems incurred in the wireless domain, the typical approach has been to fall back on conventional path finding methods and patch ant-based systems with extra mechanisms, rather than devising new approaches. As an alternative direction for future research, we suggest to instead adapt the execution of the ant-based primitives to the context of wireless networks. One such approach is OAS [19]. The idea is to apply the opportunitic forwarding paradigm [5] to the forwarding of ant agents. The fundamental changes are to defer the stochastic forwarding decision until after the reception of an ant by the potential next-hop nodes, and to execute it in a distributed manner among the receivers and in the time-domain rather than at the sender node and in the space-domain. This yields a reduction of the overhead and improves the adaptivity of the system compared to its unicast counterpart. Preliminary experiments indicate that such a system is able to achieve to efficiently spread the load along multiple paths [20].

4 Concluding Remarks

In this paper, we have retrospectively looked at 10 years of research in the field of ant-based systems for wireless networks. We have questionned the suitability of the approach for generic MANETs as well as for WSNs. We have presented the evolution and the current status of the research in this field and sketched a possible future direction of research for WMNs.

The suitability issue may also be raised for ant-based path management of dynamic telecommunication networks in general. In our view, the future might be in what we term *ant-assisted* systems rather than in ant-based systems. Rather than being the sole mean for self-organization, ant-based techniques seem more suitable as an element of a coumpound system using an array of techniques. In such a system, ants are not used for the primary path management task, but instead as a background process for the optimization of the primary system in the long run, e.g. the tuning of internal parameters, the detection of failure situations or degraded performance, or the optimization of external parameters affecting the path management problem. Applying ant-based techniques to the long-term optimization of the channel assignment in WMNs without relying on ants to solve the routing problem itself would for instance fall into this category.

References

1. Akyildiz, I., Wang, X., Wang, W.: Wireless mesh networks: a survey. Computer Networks 47(4), 445–487 (2005)
2. Akyildiz, I., Su, W., Sankarasubramaniam, Y., Cayirci, E.: Wireless sensor networks: a survey. Computer Networks 38(4), 393–422 (2002)
3. Baras, J.S., Mehta, H.: A probabilistic emergent routing algorithm for mobile ad hoc networks. In: 1st International Symposium on Modeling and Optimization in Mobile, Ad Hoc, and Wireless Networks, WiOpt (2003)
4. Bokhari, F., Zaruba, G.: AMIRA: interference-aware routing using ant colony optimization in wireless mesh networks. In: IEEE Wireless Communications and Networking Conference, WCNC (2009)
5. Bruno, R., Nurchis, M.: Survey on diversity-based routing in wireless mesh networks: Challenges and solutions. Computer Communications 33(3), 269–282 (2010)
6. Correia, F., Vazão, T.: Simple ant routing algorithm strategies for a (multipurpose) MANET model. Ad Hoc Networks 8(8), 810–823 (2010)
7. Costa, D., Hertz, A.: Ants can colour graphs. Journal of the Operational Research Society 48(3), 295–305 (1997)
8. De Couto, D., Aguayo, D., Chambers, B., Morris, R.: Performance of multihop wireless networks: Shortest path is not enough. In: 1st Workshop on Hot Topics in Networks (HotNets), ACM SIGCOMM (2002)
9. Dhillon, S.S., Arbona, X., Van Mieghem, P.: Ant routing in mobile ad hoc networks. In: 3rd IEEE International Conference on Networking and Services, ICNS (2007)
10. Di Caro, G.A., Dorigo, M.: AntNet: Distributed Stigmergetic Control for Communications Networks. Journal of Artificial Intelligence Research 9, 317–365 (1998)
11. Dorigo, M., Di Caro, G.A., Gambardella, L.M.: Ant algorithms for discrete optimization. Artificial Life 5(2), 137–172 (1999)
12. Ducatelle, F., Di Caro, G.A., Gambardella, L.M.: An Analysis of the Different Components of the AntHocNet Routing Algorithm. In: Dorigo, M., Gambardella, L.M., Birattari, M., Martinoli, A., Poli, R., Stützle, T. (eds.) ANTS 2006. LNCS, vol. 4150, pp. 37–48. Springer, Heidelberg (2006)
13. Ducatelle, F.: Adaptive Routing in Ad Hoc Wireless Multi-hop Networks. Ph.D. thesis, University of Lugano, Switzerland (2007)
14. Ducatelle, F., Di Caro, G.A., Gambardella, L.M.: A new approach for integrating proactive and reactive routing in MANETs. In: 5th IEEE International Conference on Mobile Ad Hoc and Sensor Systems, MASS (2008)
15. Ducatelle, F., Di Caro, G.A., Gambardella, L.M.: Principles and applications of swarm intelligence for adaptive routing in telecommunications networks. Swarm Intelligence 4(3), 173–198 (2010)
16. Günes, M., Sorges, U., Bouazizi, I.: ARA-the ant-colony based routing algorithm for MANETs. In: International Workshop on Ad Hoc Networking, IWAHN (2002)
17. Jacquet, P., Muhlethaler, P., Clausen, T., Laouiti, A., Qayyum, A., Viennot, L.: Optimized link state routing protocol for ad hoc networks. In: IEEE International Multi Topic Conference, INMIC (2001)

18. Farooq, M., Di Caro, G.: Routing protocols for next-generation networks inspired by collective behaviors of insect societies: An overview. In: Blum, C., Merkle, D. (eds.) Swarm Intelligence: Introduction and Applications. Natural Computing Series, pp. 101–160. Springer, Heidelberg (2008)
19. Paquereau, L., Helvik, B.E.: Opportunistic Ant-Based Path Management for Wireless Mesh Networks. In: Dorigo, M., Birattari, M., Di Caro, G.A., Doursat, R., Engelbrecht, A.P., Floreano, D., Gambardella, L.M., Groß, R., Şahin, E., Sayama, H., Stützle, T. (eds.) ANTS 2010. LNCS, vol. 6234, pp. 480–487. Springer, Heidelberg (2010)
20. Paquereau, L., Helvik, B.E.: Ant-Based Multipath Routing for Wireless Mesh Networks. In: Di Chio, C., Brabazon, A., Di Caro, G.A., Drechsler, R., Farooq, M., Grahl, J., Greenfield, G., Prins, C., Romero, J., Squillero, G., Tarantino, E., Tettamanzi, A.G.B., Urquhart, N., Uyar, A.Ş. (eds.) EvoApplications 2011, Part II. LNCS, vol. 6625, pp. 31–40. Springer, Heidelberg (2011)
21. Perkins, C., Royer, E.: Ad-hoc on-demand distance vector routing. In: 2nd IEEE Workshop on Mobile Computing and Applications (1999)
22. Saleem, M., Di Caro, G.A., Farooq, M.: A review of swarm intelligence based routing protocols for wireless sensor networks. Information Sciences 181(20), 4597–4624 (2011)
23. Shen, C.C., Jaikaeo, C.: Ad hoc multicast routing algorithm with swarm intelligence. Mobile Networks and Applications 10(1), 47–59 (2005)

Triadic Motifs and Dyadic Self-Organization in the World Trade Network

Tiziano Squartini[1,2,3] and Diego Garlaschelli[1]

[1] Instituut-Lorentz for Theoretical Physics, Leiden Institute of Physics, University of Leiden, Niels Bohrweg 2, 2333 CA Leiden, The Netherlands
[2] Department of Physics, University of Siena
[3] Center for the Study of Complex Systems, via Roma 56, 53100 Siena, Italy

Abstract. In self-organizing networks, topology and dynamics coevolve in a continuous feedback, without exogenous driving. The World Trade Network (WTN) is one of the few empirically well documented examples of self-organizing networks: its topology depends on the GDP of world countries, which in turn depends on the structure of trade. Therefore, understanding the WTN topological properties deviating from randomness provides direct empirical information about the structural effects of self-organization. Here, using an analytical pattern-detection method we have recently proposed, we study the occurrence of triadic 'motifs' (three-vertices subgraphs) in the WTN between 1950 and 2000. We find that motifs are not explained by only the in- and out-degree sequences, but they are completely explained if also the numbers of reciprocal edges are taken into account. This implies that the self-organization process underlying the evolution of the WTN is almost completely encoded into the dyadic structure, which strongly depends on reciprocity.

1 Introduction

The global economy is a prototypic example of complex self-organizing system, whose collective properties emerge spontaneously through many local interactions. In particular, international trade between countries defines a complex network which arises as the combination of many independent choices of firms. It was shown that the topology of the World Trade Network (WTN) strongly depends on the Gross Domestic Product (GDP) of world countries [1]. On the other hand, the GDP depends on international trade by definition [2], which implies that the WTN is a remarkably well documented example of adaptive network, where dynamics and topology coevolve in a continuous feedback. In general, understanding self-organizing networks is a major challenge for science, as only few models of such networks are analytically solvable [3]. However, in the particular case of the WTN, the binary topology of the network is found to be extremely well reproduced by a null model which incorporates the degree sequence [4]. These results, which have been obtained using a fast network randomization method that we have recently proposed [5], make the WTN particularly interesting. In this paper, after briefly reviewing our randomization method, we apply it

F.A. Kuipers and P.E. Heegaard (Eds.): IWSOS 2012, LNCS 7166, pp. 24–35, 2012.

to study the occurrence of triadic 'motifs', i.e. directed patterns involving three vertices (see Fig.1). We show that, unlike other properties which have been studied elsewhere [4], the occurrence of motifs is not explained by only the in- and out-degrees of vertices. However, if also the numbers of reciprocal links of each vertex (the *reciprocal degree sequence*) are taken into account, the occurrences of triadic motifs are almost completely reproduced. This implies that, if local information is enhanced in order to take into account the reciprocity structure, motifs display no significant deviations from random expectations. Therefore the (in principle complicated) self-organization process underlying the evolution of the WTN turns out to be relatively simply encoded into the local dyadic structure, which separately specifies the number of reciprocated and non-reciprocated links per vertex. Thus the dyadic structure appears to carry a large amount of information about the system.

2   Searching for Non-random Patterns in the WTN

In this section we briefly summarize our recently proposed randomization method and how it can be used to detect patterns when local constraints are considered.

2.1   Network Pattern Detection: The Randomization Method

Our method, which is based on the maximum-likelihood estimation of maximum-entropy models of graphs, introduces a family of null models of a real network and uses it to detect topological patterns analytically [5]. Defining a null model means setting up a method to assign probabilites [6,7,8]. In our approach, a real network $\mathbf{G}^*$ with N vertices is given (either a binary or a weighted graph, and either directed or undirected, whose generic entry is g_{ij}) and a way to generate a family $\mathcal{G}$ of randomized variants of $\mathbf{G}^*$ is provided, by assigning each graph $\mathbf{G} \in \mathcal{G}$ a probability $P(\mathbf{G})$. In the method, the probabilities $P(\mathbf{G})$ are such that a maximally random ensemble of networks is generated, under the constraint that, on average, a set $\{C_a\}$ of desired topological properties is set equal to the values $\{C_a(\mathbf{G}^*)\}$ observed in the real network $\mathbf{G}^*$. This is achieved as the result of a constrained Shannon-Gibbs entropy maximization [8]

$$S \equiv - \sum_{\mathbf{G}} P(\mathbf{G}) \ln P(\mathbf{G}) \tag{1}$$

and the imposed constraints are the normalization of the probability and the average values of a number of chosen properties, $\{C_a\}$:

$$1 = \sum_{\mathbf{G}} P(\mathbf{G}), \ \langle C_a \rangle \equiv \sum_{\mathbf{G}} C_a(\mathbf{G}) P(\mathbf{G}) \ . \tag{2}$$

This optimization leads to exponential probability coefficients [8]

$$P(\mathbf{G}) = \frac{e^{-H(\mathbf{G}, \boldsymbol{\theta})}}{Z} \tag{3}$$

where the linear combination of the contraints $H(\mathbf{G}, \boldsymbol{\theta}) \equiv \sum_a \theta_a C_a(\mathbf{G})$ is called *graph hamiltonian* (the coefficients $\{\theta_a\}$ are free parameters, acting as Lagrange multipliers controlling the expected values $\{\langle C_a \rangle\}$) and the denominator $Z \equiv \sum_{\mathbf{G}} e^{-H(\mathbf{G}, \boldsymbol{\theta})}$ is called *partition function*.

The next step is the maximization of the probability $P(\mathbf{G}^*)$ to obtain the observed graph $\mathbf{G}^*$, i.e. the real-world network to randomize [5]. This step fixes the values of the Lagrange multipliers as they are found by maximizing the log-likelihood

$$\mathcal{L}(\boldsymbol{\theta}) \equiv \ln P(\mathbf{G}^* | \boldsymbol{\theta}) = -H(\mathbf{G}^*, \boldsymbol{\theta}) - \ln Z(\boldsymbol{\theta}) \tag{4}$$

to obtain the real network $\mathbf{G}^*$. It can be easily verified [9] that this is achieved by the parameter values $\boldsymbol{\theta}^*$ satisfying

$$\langle C_a \rangle^* = \sum_{\mathbf{G}} C_a(\mathbf{G}) P(\mathbf{G} | \boldsymbol{\theta}^*) = C_a(\mathbf{G}^*) \quad \forall a \tag{5}$$

that is, that the ensemble average of each constraint, $\langle C_a \rangle$, equals the observed value on the real network, $C_a(\mathbf{G}^*)$. Once the numerical values of the Lagrange multipliers are found, they can be used to find the ensemble average $\langle X \rangle^*$ of any topological property X of interest [5]:

$$\langle X \rangle^* = \sum_{\mathbf{G}} X(\mathbf{G}) P(\mathbf{G} | \boldsymbol{\theta}^*) \ . \tag{6}$$

The exact computation of the expected values can be very difficult. For this reason it is often necessary to rest on the *linear approximation method* [5]. However, in the present study we will consider particular topological properties X (i.e. motif counts, see below) whose expected value can be evaluated *exactly*. Our method also allows to obtain the variance of X by applying the usual definition:

$$\sigma^2[X] = \langle [X(\mathbf{G}) - \langle X \rangle)]^2 \rangle = \sum_{i,j} \sum_{t,s} \sigma[g_{ij}, g_{ts}] \left(\frac{\partial X}{\partial g_{ij}} \frac{\partial X}{\partial g_{ts}} \right)_{\mathbf{G} = \langle \mathbf{G} \rangle} \tag{7}$$

where $\sigma[g_{ij}, g_{ts}]$ is the covariance of the adjacency matrix elements g_{ij} and g_{ts}. This formula can be greatly simplified by considering probabilities that are factorizable in terms of dyadic probabilities, as follows

$$P(\mathbf{G} | \boldsymbol{\theta}) = \prod_{i<j} D_{ij}(g_{ij}, g_{ji} | \boldsymbol{\theta}) \tag{8}$$

where the product runs over all the dyads, that is the unordered pairs of vertices (i, j) (with $i < j$), and $D_{ij}(g, g' | \boldsymbol{\theta})$ is the joint probability that $g_{ij} = g$ and $g_{ji} = g'$. Finally, the variance of X evaluated in $\boldsymbol{\theta}^*$ becomes

$$(\sigma^*[X])^2 = \sum_{i,j} \left[\left(\sigma^*[g_{ij}] \frac{\partial X}{\partial g_{ij}} \right)^2_{\mathbf{G} = \langle \mathbf{G} \rangle^*} + \sigma^*[g_{ij}, g_{ji}] \left(\frac{\partial X}{\partial g_{ij}} \frac{\partial X}{\partial g_{ji}} \right)_{\mathbf{G} = \langle \mathbf{G} \rangle^*} \right] \ . \tag{9}$$

The joint knowledge of $\langle X \rangle^*$ and $\sigma^*[X]$ allows to detect deviations from randomness in the observed topology. In particular, as we show later, it is possible to calculate by how many standard deviations the observed value X^* differs from the expected value $\langle X \rangle^*$. Quantities which are consistent with their expected value are explained by the enforced constraints $\{C_a\}$. On the other hand, significantly deviating properies cannot be traced back to the constraints and therefore signal the incompleteness of the information encoded in the constraints. Other approaches achieve this result by explicitly generating many randomized variants of the real network, measuring X on each such variant, and finally computing the sample average and standard deviation of X [6]. This is extremely time consuming, especially for complicated topological properties. By contrast, our method is entirely analytical. It yields any expected quantity $\langle X \rangle^*$ in a time as short as that required in order to measure X^* on the single network $\mathbf{G}^*$ [5].

2.2 The Role of Local Constraints

If the network is a binary graph (i.e. if each graph $\mathbf{G}$ in the ensemble is uniquely specified by its adjacency matrix $\mathbf{A}$), then the simplest (i.e. local) choice of the constraints is the *degree sequence*, i.e. the vector of degrees (numbers of incident links) of all vertices. For directed networks, which are our interest here, there are actually two degree sequences: the observed in-degree sequence $\mathbf{k}^{in}(\mathbf{A}^*)$ (with $k_i^{in} = \sum_{j \neq i} a_{ji}$) and the observed out-degree sequence $\mathbf{k}^{out}(\mathbf{A}^*)$ (with $k_i^{out} = \sum_{j \neq i} a_{ij}$). This null model, which is known as the *directed configuration model* (DCM), can be completely dealt with analytically using our method (see Appendix). When applied to the WTN, the DCM shows that many topological properties (such as the degree-degree correlations and the directed clustering coefficients) are in complete accordance with the expectations [4]. This shows that the degree sequences $\mathbf{k}^{in}$ and $\mathbf{k}^{out}$ are extremely informative, as their (partial) knowledge allows to reconstruct many aspects of the (complete) topology. On the other hand, it was also shown that the *reciprocity* of the WTN is highly non-trivial [10]. This means that the occurrence of reciprocal links is much higher than expected under any model which, as the DCM, treats two reciprocal links (e.g. $i \to j$ and $j \to i$) as statistically independent. A direct consequence is that the reciprocity, as well as any higher-order directional pattern, should not be reproduced by the DCM. These seemingly conflicting results can only be reconciled if, for some reason, the topological properties that have been studied under the DCM [4] mask the effects of reciprocity. In particular, the directed clustering coefficients, which are based on ratios of realized triangles over the maximum number for each vertex, may show no overall deviation from the DCM, even if the numerator and denominator separately deviate from it. In what follows, we investigate this possibility by considering all the observed subgraphs of three vertices (which include both open and closed triangles) separately. Also, we will use an additional null model which also takes the number of reciprocal links of each vertex into account. This second null model is the *reciprocal configuration model* (RCM) [5,10,11]. The local constraints defining it are the three, observed directed-degree sequences $\mathbf{k}^{\rightarrow}(\mathbf{A}^*)$, with $k_i^{\rightarrow} \equiv \sum_{j \neq i} a_{ij}^{\rightarrow}$

(non-reciprocated out degree), $\mathbf{k}^{\leftarrow}(\mathbf{A}^*)$, with $k_i^{\leftarrow} \equiv \sum_{j \neq i} a_{ij}^{\leftarrow}$ (non-reciprocated in-degree) and $\mathbf{k}^{\leftrightarrow}(\mathbf{A}^*)\}$, with $k_i^{\leftrightarrow} \equiv \sum_{j \neq i} a_{ij}^{\leftrightarrow}$ (reciprocated degree) to be imposed across the ensemble of networks having the same number of vertices of the observed configuration and, on average, the above-mentioned directed-degree sequences. In the Appendix we describe both the DCM and the RCM in more detail, and derive their expectations explicitly.

2.3   Triadic Motifs in the WTN

In the following analyses, we use yearly bilateral data on exports and imports from the Gleditsch Database[1] to analyse the six years 1950, 1960, 1970, 1980, 1990, 2000. This database contains aggregated trade data between countries, i.e. data as they result by summing the single commodity-specific trade exchanges. So we end up with six different, real, asymmetric matrices with entries $m_{ij}^{agg}(y)$ ($y = 1950, 1960 \ldots 2000$). These adjacency matrix elements are the fundamental data allowing us to obtain all the possible representations of the WTN: to build the binary, directed representation we are interested in here, we restrict ourselves to consider two different vertices as linked, whenever the corresponding element $m_{ij}^{agg}(y)$ is strictly positive. This implies that the adjacency matrix of the binary, directed representation of the WTN in year y is simply obtained by applying the Heaviside step function to the database entries, i.e. $a_{ij}(y) = \Theta[m_{ij}^{agg}(y)]$.

Triadic motifs, i.e. all the possibile directed patterns connecting three vertices, are the natural generalizations of directed clustering coefficients [12] and the starting point for the understanding of a complex network's self-organization in communities. Thirteen, non-isomorphic, triadic directed patterns can be indentified and classified [13]. Given a real, binary, directed matrix $\mathbf{A}^*$, the motifs occurrences N_m can be written in at least two different ways (see Table 1). The first one prescribes to define them in terms of the adjacency matrix entries, $\{a_{ij}\}$. The second one allows to compactly express the coefficients N_m by introducing the following dyadic variables

$$a_{ij}^{\rightarrow} \equiv a_{ij}(1 - a_{ji}), \; a_{ij}^{\leftarrow} \equiv a_{ji}(1 - a_{ij}), \; a_{ij}^{\leftrightarrow} \equiv a_{ij}a_{ji}, \; a_{ij}^{\nleftrightarrow} \equiv (1 - a_{ij})(1 - a_{ji}) \quad (10)$$

thus making the role of reciprocity explicit. However, the number of occurrences of the particular motif m (where m ranges from 1 to 13) as measured on the observed network $\mathbf{A}^*$ is uninformative unless a comparison with a properly defined null model is made (see Appendix). This implies that the occurrence of a motif should be compared with its expected value $\langle N_m \rangle^*$, as computed under the chosen null model. This can be compactly achieved by introducing the so-called z-score

$$z[N_m] \equiv \frac{N_m(\mathbf{A}^*) - \langle N_m \rangle^*}{\sigma^*[N_m]} \quad (11)$$

[1] Gleditsch, K.S.: Expanded trade and GDP data. Jour. Confl. Res. **46** (2002) 712–724.

Fig. 1. The triadic, binary, directed motifs

Table 1. Classification (after [13]) and definitions of the triadic motifs

Motif m N_m: 1st definition	N_m: 2nd definition
1 $\sum_{i\neq j\neq k}(1-a_{ij})a_{ji}a_{jk}(1-a_{kj})(1-a_{ik})(1-a_{ki})$	$\sum_{i\neq j\neq k} a_{ij}^{\leftarrow} a_{jk}^{\rightarrow} a_{ik}^{\leftrightarrow}$
2 $\sum_{i\neq j\neq k} a_{ij}(1-a_{ji})a_{jk}(1-a_{kj})(1-a_{ik})(1-a_{ki})$	$\sum_{i\neq j\neq k} a_{ij}^{\rightarrow} a_{jk}^{\rightarrow} a_{ik}^{\leftrightarrow}$
3 $\sum_{i\neq j\neq k} a_{ij}a_{ji}a_{jk}(1-a_{kj})(1-a_{ik})(1-a_{ki})$	$\sum_{i\neq j\neq k} a_{ij}^{\leftrightarrow} a_{jk}^{\rightarrow} a_{ik}^{\leftrightarrow}$
4 $\sum_{i\neq j\neq k}(1-a_{ij})(1-a_{ji})a_{jk}(1-a_{kj})a_{ik}(1-a_{ki})$	$\sum_{i\neq j\neq k} a_{ij}^{\leftrightarrow} a_{jk}^{\rightarrow} a_{ik}^{\leftarrow}$
5 $\sum_{i\neq j\neq k}(1-a_{ij})a_{ji}a_{jk}(1-a_{kj})a_{ik}(1-a_{ki})$	$\sum_{i\neq j\neq k} a_{ij}^{\leftarrow} a_{jk}^{\rightarrow} a_{ik}^{\rightarrow}$
6 $\sum_{i\neq j\neq k} a_{ij}a_{ji}a_{jk}(1-a_{kj})a_{ik}(1-a_{ki})$	$\sum_{i\neq j\neq k} a_{ij}^{\leftrightarrow} a_{jk}^{\rightarrow} a_{ik}^{\rightarrow}$
7 $\sum_{i\neq j\neq k} a_{ij}a_{ji}(1-a_{jk})a_{kj}(1-a_{ik})(1-a_{ki})$	$\sum_{i\neq j\neq k} a_{ij}^{\leftrightarrow} a_{jk}^{\leftarrow} a_{ik}^{\leftrightarrow}$
8 $\sum_{i\neq j\neq k} a_{ij}a_{ji}a_{jk}a_{kj}(1-a_{ik})(1-a_{ki})$	$\sum_{i\neq j\neq k} a_{ij}^{\leftrightarrow} a_{jk}^{\leftrightarrow} a_{ik}^{\leftrightarrow}$
9 $\sum_{i\neq j\neq k}(1-a_{ij})a_{ji}(1-a_{jk})a_{kj}a_{ik}(1-a_{ki})$	$\sum_{i\neq j\neq k} a_{ij}^{\leftarrow} a_{jk}^{\leftarrow} a_{ik}^{\rightarrow}$
10 $\sum_{i\neq j\neq k}(1-a_{ij})a_{ji}a_{jk}a_{kj}a_{ik}(1-a_{ki})$	$\sum_{i\neq j\neq k} a_{ij}^{\leftarrow} a_{jk}^{\leftrightarrow} a_{ik}^{\rightarrow}$
11 $\sum_{i\neq j\neq k} a_{ij}(1-a_{ji})a_{jk}a_{kj}a_{ik}(1-a_{ki})$	$\sum_{i\neq j\neq k} a_{ij}^{\rightarrow} a_{jk}^{\leftrightarrow} a_{ik}^{\rightarrow}$
12 $\sum_{i\neq j\neq k} a_{ij}a_{ji}a_{jk}a_{kj}a_{ik}(1-a_{ki})$	$\sum_{i\neq j\neq k} a_{ij}^{\leftrightarrow} a_{jk}^{\leftrightarrow} a_{ik}^{\rightarrow}$
13 $\sum_{i\neq j\neq k} a_{ij}a_{ji}a_{jk}a_{kj}a_{ik}a_{ki}$	$\sum_{i\neq j\neq k} a_{ij}^{\leftrightarrow} a_{jk}^{\leftrightarrow} a_{ik}^{\leftrightarrow}$

measuring by how many standard deviations, $\sigma^*[N_m]$, the observed and the expected occurrences of motif m differ. Large absolute values of $z[N_m]$ indicate motifs that are either over- or under-represented under the particular null model considered and therefore not explained by the constraints defining it, as shown in Fig.2 and Fig.3 and discussed in the next section.

3 Results and Discussion

Fig.2 and Fig.3 show the z-scores for all the 13, triadic, binary, directed motifs, computed for the six different snapshots of the World Trade Network corresponding to the decades 1950, 1960, 1970, 1980, 1990, 2000, for both the DCM and the RCM. We also show the six lines $z=\pm1$, $z=\pm2$ and $z=\pm3$ to highlight the region within 3 sigmas from the expectation value. The analysis reveals a dramatic difference between the predictions of the two null models. The presence of intrinsically directed trading relationships implies that reciprocity is a fundamental quantity, shaping the network of exchanges among world-countries.

The reciprocity of the WTN is known to be very high [10] and this has strong effects on its motif structure. This implies that, in order to reproduce the topology of the network, it is essential to reproduce its dyadic structure, which in turn

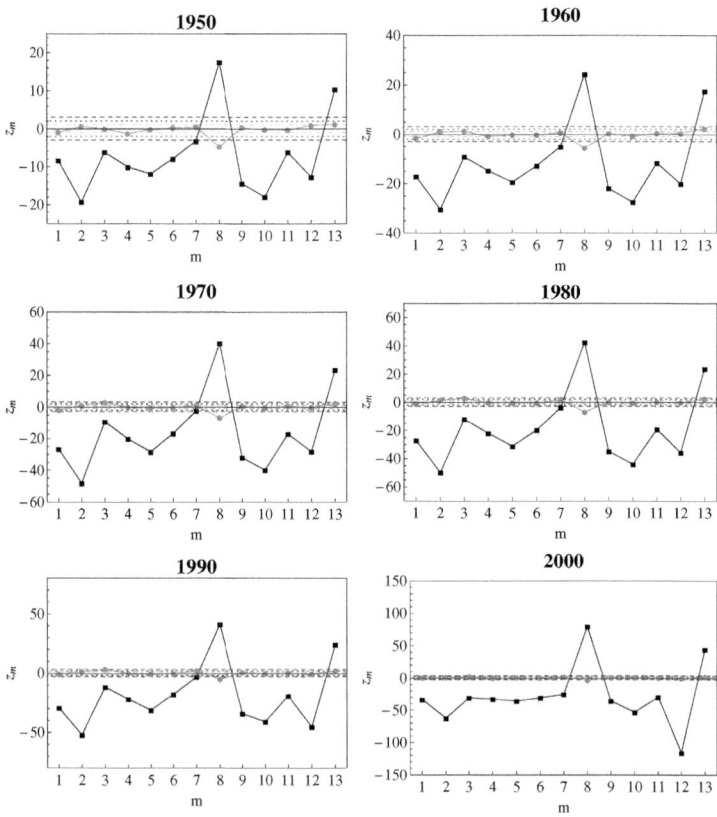

Fig. 2. z-scores of the 13 triadic, binary, directed motifs for the six decades of the WTN, under the DCM (■) and under the RCM (•). The dashed, red lines represent the values $z = \pm 3$, the dotted, purple lines the values $z = \pm 2$ and the dot-dashed, pink lines the values $z = \pm 1$.

strongly depends on reciprocity. This also implies that the RCM should fluctuate less than the DCM, because of this huge amount of additional information. Indeed, this is confirmed by our analysis.

In fact, we find that, unlike other topological properties [4], triadic motifs are systematically not reproduced if only the in-degree and out-degree sequences are taken into account. In particular, the observed motifs counts deviate by up to 100 standard deviations from the expected ones. By contrast, if also the reciprocity (which separately specifies the number of reciprocated links) is introduced in the null model, triadic motifs are almost always consistent with expectations within one standard deviation, as for $m = 5, 10, 11, 12$ or at most two standard deviations, as $m = 1, 4$. Moreover, the significance profiles are almost completely inverted: some of the motifs being over(under)-represented without the reciprocity constraint, as $m = 8$ ($m = 2$), become under(over)-represented with the reciprocity constraint.

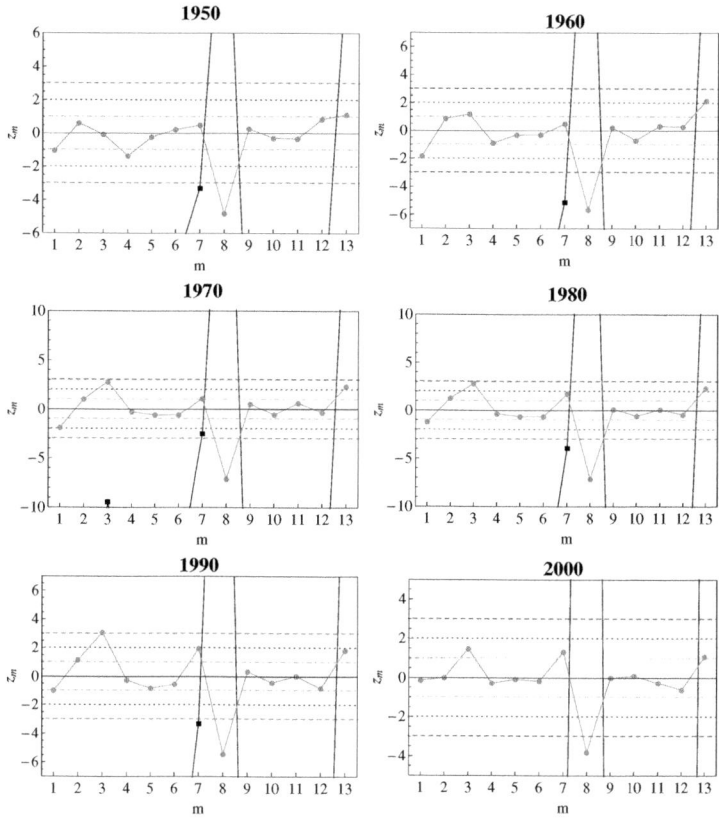

Fig. 3. z-scores of the 13 triadic, binary, directed motifs for the six decades of the WTN, under the DCM (■) and under the RCM (●): zoom of Fig. 2. The dashed, red lines represent the values $z = \pm 3$, the dotted, purple lines the values $z = \pm 2$ and the dot-dashed, pink lines the values $z = \pm 1$.

4 Conclusions

The WTN is a particularly interesting network which is known to be driven by a self-organization process involving the global economy. Empirically, it is a very well documented network which allows to test predictions of null models about the key ingredients shaping its topology. In this paper, we aimed at isolating the key properties of the WTN topology where the effects of self-organizations can be clearly detected as deviations from randomness. While the DCM, which takes into account the number of incoming and outgoing connections of all vertices, reproduces many other topological properties, it cannot replicate the observed triadic motifs. On the other hand, we found that the RCM, which also takes into account the numbers of reciprocated links, can replicate almost perfectly the triadic structure. We therefore found that the process underlying the evolution of the WTN is mainly encoded into the dyadic structure, which carries a large amount of information about the system: so, the upgrade to the RCM is necessary and the possibility to treat the RCM analytically using our method is therefore an important step forward. The result that dyadic properties almost completely explain triadic ones suggests that in the WTN also higher-order properties (e.g. subgraphs of four or more vertices, and in general even the existence of denser communities of vertices) will be explained mostly by dyadic properties. This raises the important question whether the available community detection methods are successful in identifying communities which are not explained by local constraints. In particular, our results suggest that modularity-based community detection methods will detect communities in the WTN if a null model without reciprocity (DCM) is used, while they should find weak or no community structure if a null model including reciprocity (RCM) is used. In both cases, as we have already claimed [5], the available expressions for the modularity (which make the strong assumption of sparse networks) should be revised using the correct expectation values provided by our method, since the WTN is a very dense network. We will address these issues in future work.

Acknowledgements. This work was supported by the Dutch Econophysics Foundation (Stichting Econophysics, Leiden, The Netherlands) with funds from beneficiaries of Duyfken Trading Knowledge BV, Amsterdam, The Netherlands.

Appendix

The Directed Configuration Model

Given a real, binary, directed network $\mathbf{A}^*$ with out-degree and in-degree sequences $\mathbf{k}^{out}(\mathbf{A}^*)$ and $\mathbf{k}^{in}(\mathbf{A}^*)$, the method described in section 2.1 can be specified in the following way.

The DCM Hamiltonian. The Hamiltonian implementing the DCM is

$$H(\mathbf{A}, \boldsymbol{\alpha}, \boldsymbol{\beta}) = \sum_i [\alpha_i k_i^{out}(\mathbf{A}) + \beta_i k_i^{in}(\mathbf{A})] = \sum_{i \neq j} (\alpha_i + \beta_j) a_{ij} \; ; \qquad (12)$$

the partition function can be calculated as in [8], $Z(\boldsymbol{\alpha}, \boldsymbol{\beta}) = \sum_{\mathbf{A}} e^{-H(\mathbf{A}, \boldsymbol{\alpha}, \boldsymbol{\beta})} = \prod_{i \neq j} (1 + e^{-\alpha_i - \beta_j})$, and the graph probability is

$$P(\mathbf{A}|\boldsymbol{\alpha}, \boldsymbol{\beta}) \equiv \prod_{i < j} D_{ij}(a_{ij}, a_{ji}|\alpha_i, \alpha_j, \beta_i, \beta_j) = \prod_{i \neq j} p_{ij}^{a_{ij}} (1 - p_{ij})^{(1 - a_{ij})} \qquad (13)$$

and, by setting $x_i \equiv e^{-\alpha_i}$ and $y_i \equiv e^{-\beta_i}$,

$$p_{ij} = \frac{e^{-\alpha_i - \beta_j}}{1 + e^{-\alpha_i - \beta_j}} \equiv \frac{x_i y_j}{1 + x_i y_j} \; . \qquad (14)$$

The Log-Likelihood Equations. The log-likelihood function to maximize is

$$\mathcal{L}(\mathbf{x},\mathbf{y}) = \sum_i \left[k_i^{\text{out}}(\mathbf{A}^*) \ln x_i + k_i^{\text{in}}(\mathbf{A}^*) \ln y_i \right] - \sum_{i \neq j} \ln(1 + x_i y_j) \qquad (15)$$

and the values $\mathbf{x}^*$, $\mathbf{y}^*$ corresponding to the point of maximum can be found by solving the following system

$$\begin{cases} k_i^{\text{in}}(\mathbf{A}^*) = \sum_{j \neq i} \frac{x_i^* y_j^*}{1 + x_i^* y_j^*} & \forall i \\ k_i^{\text{out}}(\mathbf{A}^*) = \sum_{j \neq i} \frac{x_j^* y_i^*}{1 + x_j^* y_i^*} & \forall i \end{cases} \qquad (16)$$

Expectation Values and Variances. The expectation value and variance of a_{ij} are, respectively, $\langle a_{ij} \rangle^* = p_{ij}^*$ and $(\sigma^*[a_{ij}])^2 = \langle a_{ij} \rangle^*(1 - \langle a_{ij} \rangle^*) = p_{ij}^*(1 - p_{ij}^*)$. Distinct pairs of vertices are independent random variables; since the first definition of the motifs occurrences only involves products of the adjacency matrix entries, their expectation values can be easily computed, as shown in Table 2. The variance of N_m becomes

$$(\sigma^*[N_m])^2 = \sum_{i,j} \left[\left(\sigma^*[a_{ij}] \frac{\partial N_m}{\partial a_{ij}} \right)^2_{\mathbf{A} = \langle \mathbf{A} \rangle^*} \right] \qquad (17)$$

considering that $\sigma^*[a_{ij}, a_{ji}] = \langle a_{ij} a_{ji} \rangle^* - \langle a_{ij} \rangle^* \langle a_{ji} \rangle^* = 0$.

The Reciprocal Configuration Model

Given a real binary directed network $\mathbf{A}^*$ with directed-degree sequences $\mathbf{k}^{\rightarrow}(\mathbf{A}^*)$, $\mathbf{k}^{\leftarrow}(\mathbf{A}^*)$ and $\mathbf{k}^{\leftrightarrow}(\mathbf{A}^*)$ where

$$k_i^{\rightarrow}(\mathbf{A}^*) \equiv \sum_{j \neq i} a_{ij}^*(1 - a_{ji}^*), \ k_i^{\leftarrow}(\mathbf{A}^*) \equiv \sum_{j \neq i} a_{ji}^*(1 - a_{ij}^*), \ k_i^{\leftrightarrow}(\mathbf{A}^*) \equiv \sum_{j \neq i} a_{ij}^* a_{ji}^*$$

$$(18)$$

the randomization procedure can be specified in the following way.

The RCM Hamiltonian. The Hamiltonian implementing the RCM is

$$H(\mathbf{A}, \boldsymbol{\alpha}, \boldsymbol{\beta}, \boldsymbol{\gamma}) = \sum_i [\alpha_i k_i^{\rightarrow}(\mathbf{A}) + \beta_i k_i^{\leftarrow}(\mathbf{A}) + \gamma_i k_i^{\leftrightarrow}(\mathbf{A})] \ , \qquad (19)$$

the partition function can be calculated as in [11], $Z(\boldsymbol{\alpha}, \boldsymbol{\beta}, \boldsymbol{\gamma}) = \prod_{i<j}(1 + e^{-\alpha_i - \beta_j} + e^{-\alpha_j - \beta_i} + e^{-\gamma_i - \gamma_j})$, and the graph probability is

$$P(\mathbf{A}|\boldsymbol{\alpha}, \boldsymbol{\beta}, \boldsymbol{\gamma}) = \prod_{i<j} D_{ij}(a_{ij}, a_{ji}|\alpha_i, \alpha_j, \beta_i, \beta_j, \gamma_i, \gamma_j)$$

$$= \prod_{i<j} (p_{ij}^{\rightarrow})^{a_{ij}^{\rightarrow}} (p_{ij}^{\leftarrow})^{a_{ij}^{\leftarrow}} (p_{ij}^{\leftrightarrow})^{a_{ij}^{\leftrightarrow}} (p_{ij}^{\nleftrightarrow})^{a_{ij}^{\nleftrightarrow}} \qquad (20)$$

Table 2. Expectation values of the triadic motifs

Motif m	$\langle N_m \rangle_{DCM}$	$\langle N_m \rangle_{RCM}$
1	$\sum_{i \neq j \neq k}(1-p_{ij})p_{ji}p_{jk}(1-p_{kj})(1-p_{ik})(1-p_{ki})$	$\sum_{i \neq j \neq k} p_{ij}^{\leftarrow} p_{jk}^{\rightarrow} p_{ik}^{\leftrightarrow}$
2	$\sum_{i \neq j \neq k}p_{ij}(1-p_{ji})p_{jk}(1-p_{kj})(1-p_{ik})(1-p_{ki})$	$\sum_{i \neq j \neq k} p_{ij}^{\rightarrow} p_{jk}^{\rightarrow} p_{ik}^{\leftrightarrow}$
3	$\sum_{i \neq j \neq k}p_{ij}p_{ji}p_{jk}(1-p_{kj})(1-p_{ik})(1-p_{ki})$	$\sum_{i \neq j \neq k} p_{ij}^{\leftrightarrow} p_{jk}^{\rightarrow} p_{ik}^{\leftrightarrow}$
4	$\sum_{i \neq j \neq k}(1-p_{ij})(1-p_{ji})p_{jk}(1-p_{kj})p_{ik}(1-p_{ki})$	$\sum_{i \neq j \neq k} p_{ij}^{\nleftrightarrow} p_{jk}^{\rightarrow} p_{ik}^{\rightarrow}$
5	$\sum_{i \neq j \neq k}(1-p_{ij})p_{ji}p_{jk}(1-p_{kj})p_{ik}(1-p_{ki})$	$\sum_{i \neq j \neq k} p_{ij}^{\leftarrow} p_{jk}^{\rightarrow} p_{ik}^{\rightarrow}$
6	$\sum_{i \neq j \neq k}p_{ij}p_{ji}p_{jk}(1-p_{kj})p_{ik}(1-p_{ki})$	$\sum_{i \neq j \neq k} p_{ij}^{\leftrightarrow} p_{jk}^{\rightarrow} p_{ik}^{\rightarrow}$
7	$\sum_{i \neq j \neq k}p_{ij}p_{ji}(1-p_{jk})p_{kj}(1-p_{ik})(1-p_{ki})$	$\sum_{i \neq j \neq k} p_{ij}^{\leftrightarrow} p_{jk}^{\leftarrow} p_{ik}^{\leftrightarrow}$
8	$\sum_{i \neq j \neq k}p_{ij}p_{ji}p_{jk}p_{kj}(1-p_{ik})(1-p_{ki})$	$\sum_{i \neq j \neq k} p_{ij}^{\leftrightarrow} p_{jk}^{\leftrightarrow} p_{ik}^{\leftrightarrow}$
9	$\sum_{i \neq j \neq k}(1-p_{ij})p_{ji}(1-p_{jk})p_{kj}p_{ik}(1-p_{ki})$	$\sum_{i \neq j \neq k} p_{ij}^{\leftarrow} p_{jk}^{\leftarrow} p_{ik}^{\rightarrow}$
10	$\sum_{i \neq j \neq k}(1-p_{ij})p_{ji}p_{jk}p_{kj}p_{ik}(1-p_{ki})$	$\sum_{i \neq j \neq k} p_{ij}^{\leftarrow} p_{jk}^{\leftrightarrow} p_{ik}^{\rightarrow}$
11	$\sum_{i \neq j \neq k}p_{ij}(1-p_{ji})p_{jk}p_{kj}p_{ik}(1-p_{ki})$	$\sum_{i \neq j \neq k} p_{ij}^{\rightarrow} p_{jk}^{\leftrightarrow} p_{ik}^{\rightarrow}$
12	$\sum_{i \neq j \neq k}p_{ij}p_{ji}p_{jk}p_{kj}p_{ik}(1-p_{ki})$	$\sum_{i \neq j \neq k} p_{ij}^{\leftrightarrow} p_{jk}^{\leftrightarrow} p_{ik}^{\rightarrow}$
13	$\sum_{i \neq j \neq k}p_{ij}p_{ji}p_{jk}p_{kj}p_{ik}p_{ki}$	$\sum_{i \neq j \neq k} p_{ij}^{\leftrightarrow} p_{jk}^{\leftrightarrow} p_{ik}^{\leftrightarrow}$

and, by setting $x_i \equiv e^{-\alpha_i}$, $y_i \equiv e^{-\beta_i}$ and $z_i \equiv e^{-\gamma_i}$ [11],

$$p_{ij}^{\rightarrow} \equiv \frac{x_i y_j}{1+x_i y_j + x_j y_i + z_i z_j}, \quad p_{ij}^{\leftarrow} \equiv \frac{x_j y_i}{1+x_i y_j + x_j y_i + z_i z_j} \tag{21}$$

$$p_{ij}^{\leftrightarrow} \equiv \frac{z_i z_j}{1+x_i y_j + x_j y_i + z_i z_j}, \quad p_{ij}^{\nleftrightarrow} \equiv \frac{1}{1+x_i y_j + x_j y_i + z_i z_j} \tag{22}$$

The Log-Likelihood Equations. The log-likelihood function to maximize is

$$\mathcal{L}(\mathbf{x}, \mathbf{y}, \mathbf{z}) = \sum_i \left[k_i^{\rightarrow}(\mathbf{A}^*) \ln x_i + k_i^{\leftarrow}(\mathbf{A}^*) \ln y_i + k_i^{\leftrightarrow}(\mathbf{A}^*) \ln z_i \right]$$
$$- \sum_{i<j} \ln(1 + x_i y_j + x_j y_i + z_i z_j) \tag{23}$$

and the values $\mathbf{x}^*$, $\mathbf{y}^*$, $\mathbf{z}^*$ corresponding to the point of maximum can be found by solving the following system

$$\begin{cases} k_i^{\rightarrow}(\mathbf{A}^*) = \sum_{j \neq i} \frac{x_i^* y_j^*}{1+x_i^* y_j^*+x_j^* y_i^*+z_i^* z_j^*} & \forall i \\ k_i^{\leftarrow}(\mathbf{A}^*) = \sum_{j \neq i} \frac{x_j^* y_i^*}{1+x_i^* y_j^*+x_j^* y_i^*+z_i^* z_j^*} & \forall i \\ k_i^{\leftrightarrow}(\mathbf{A}^*) = \sum_{j \neq i} \frac{z_i^* z_j^*}{1+x_i^* y_j^*+x_j^* y_i^*+z_i^* z_j^*} & \forall i \end{cases} \tag{24}$$

Expectation Values and Variances. The expectation value and variance of $a_{ij}^{\rightarrow}$ (and equivalently for the other dyadic variables) are, respectively, $\langle a_{ij}^{\rightarrow} \rangle^* = (p_{ij}^{\rightarrow})^*$ and $(\sigma^*[a_{ij}^{\rightarrow}])^2 = \langle a_{ij}^{\rightarrow} \rangle^*(1-\langle a_{ij}^{\rightarrow} \rangle^*) = (p_{ij}^{\rightarrow})^*(1-(p_{ij}^{\rightarrow})^*)$. Considering that two distinct dyads can be treated as independent random variables and that the second definition of the motifs occurrences only involves products of dyads, their expectation values can be easily computed, as shown in Table 2. The variance of N_m becomes

$$(\sigma^*[N_m])^2 = \sum_{i,j} \left[\left(\sigma^*[a_{ij}] \frac{\partial N_m}{\partial a_{ij}} \right)^2_{\mathbf{A}=\langle \mathbf{A} \rangle^*} + \sigma^*[a_{ij}, a_{ji}] \left(\frac{\partial N_m}{\partial a_{ij}} \frac{\partial N_m}{\partial a_{ji}} \right)_{\mathbf{A}=\langle \mathbf{A} \rangle^*} \right]$$

(25)

where now $(\sigma^*[a_{ij}])^2 = \langle a_{ij}^{\leftrightarrow} + a_{ij}^{\rightarrow} \rangle^* (1 - \langle a_{ij}^{\leftrightarrow} + a_{ij}^{\rightarrow} \rangle^*)$ and $\sigma^*[a_{ij}, a_{ji}] = \langle a_{ij}^{\leftrightarrow} \rangle^* - \langle a_{ij}^{\leftrightarrow} + a_{ij}^{\rightarrow} \rangle^* \langle a_{ji}^{\leftrightarrow} + a_{ji}^{\rightarrow} \rangle^*$.

References

1. Garlaschelli, D., Loffredo, M.I.: Fitness-dependent topological properties of the World Trade Web. Phys. Rev. Lett. 93, 188701 (2004)
2. Garlaschelli, D., Di Matteo, T., Aste, T., Caldarelli, G., Loffredo, M.I.: Interplay between topology and dynamics in the World Trade Web. Eur. Phys. J. B 57, 159–164 (2007)
3. Garlaschelli, D., Capocci, A., Caldarelli, G.: Self-organized network evolution coupled to extremal dynamics. Nat. Phys. 3, 813–817 (2007)
4. Squartini, T., Fagiolo, G., Garlaschelli, D.: Randomizing world trade. I. A binary network analysis. Phys. Rev. E 84, 046117 (2011)
5. Squartini, T., Garlaschelli, D.: Analytical maximum-likelihood method to detect patterns in real networks. New J. Phys. 13, 083001 (2011)
6. Maslov, S., Sneppen, K., Zaliznyak, A.: Detection of topological patterns in complex networks: correlation profile of the Internet. Physica A 333, 529–540 (2004)
7. Holland, P., Leinhardt, S.: Sociological Methodology, pp. 1–45. Jossey-Bass, San Francisco (1975)
8. Park, J., Newman, M.E.J.: Statistical mechanics of networks. Phys. Rev. E 70, 066117 (2004)
9. Garlaschelli, D., Loffredo, M.I.: Maximum likelihood: extracting unbiased information from complex networks. Phys. Rev. E 78, 015101(R) (2008)
10. Garlaschelli, D., Loffredo, M.I.: Patterns of link reciprocity in directed networks. Phys. Rev. Lett. 93, 268701 (2004)
11. Garlaschelli, D., Loffredo, M.I.: Multispecies grand-canonical models for networks with reciprocity. Phys. Rev. E 73, 015101(R) (2004)
12. Fagiolo, G.: Clustering in complex directed networks. Phys. Rev. E 76, 026107 (2007)
13. Caldarelli, G.: Scale-free Networks. Complex Webs in Nature and Technology, pp. 35–38. Oxford University Press, Oxford (2007)

On Measurement of Internal Variables of Complex Self-Organized Systems and Their Relation to Multifractal Spectra

Dalibor Štys[1], Petr Jizba[2], Štěpán Papáček[1],
Tomáš Náhlík[1], and Petr Císař[1]

[1] School of Complex Sytems, Faculty of Fishery and Water Protection,
University of South Bohemia, Zámek 136, 373 33 Nové Hrady, Czech Republic
{stys,papacek,nahlik,cisar}@jcu.cz
[2] FNSPE, Czech Technical University in Prague,
Břehová 7, 115 19 Prague, Czech Republic
p.jizba@fjfi.cvut.cz

Abstract. We propose a method for characterizing structured, experimentally observable, complex self-organized systems. The method in question is based on the observation that number of self-organized systems can be mathematically perceived as consisting of several interconnected multifractal components. We illustrate our key results with ensuing applications. The relation of the results obtained to known examples of strange attractors is also discussed.

Keywords: Rényi entropy, principal component analysis, generalized dimensions, multifractal spectra.

1 Introduction

Self-organized (SO) systems typically refer to physical and biological systems in which patterns or structures present at the macroscopic level arise solely from a coherent interaction among lower level components of the system. The best known examples are undoubtedly living organisms which range in their complexity from simple cells through herds, flocks, insect colonies to humans and their herding behaviour or cities. There are many models used to generate analogies of specific features of such observable structures. The comparison of these models to observed dynamical systems is the question which awaits yet the answer. It is purpose of this paper to point out that some signatures that are met in analysis of SO systems may be inferred by the statistical approach based on a multivariate analysis. This approach has at the moment a widely utilized version epitomized by the principal component analysis (PCA) alongside with many new developments including, e.g., analyses developed for data exhibiting distribution skewness [1]. It should be, however born in mind that many realistic intensity distributions found experimentally in images of SO systems such as Belousov–Zhabotinsky reaction (see Fig. 1) or living cells (see Fig. 2) are rather complex

F.A. Kuipers and P.E. Heegaard (Eds.): IWSOS 2012, LNCS 7166, pp. 36–47, 2012.

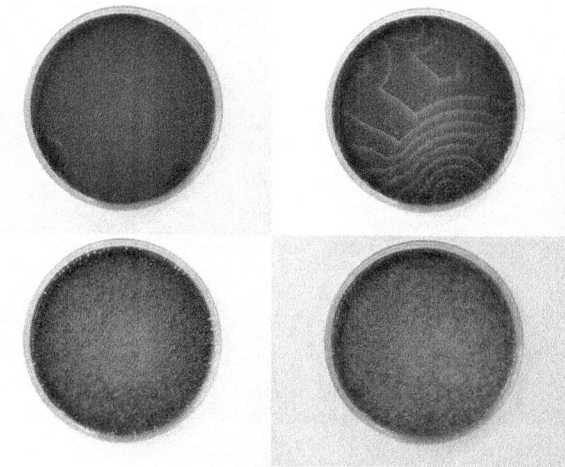

Fig. 1. Examples of structures observed in the course of Belousov–Zhabotinsky reaction. From the upper right to bottom left we depict images of 38th, 140th, 277th and 425th minute of the reaction trend.

and cannot be mapped on a simple (skewed) normal distribution. Thus, the question of multivariate analysis on data-sets coming from SO systems remains open, regardless of its wide utility, for instance, in various fields of biology and medicine [2].

In this paper we describe a surprisingly successful application of the PCA for charting the state space of the Belousov–Zhabotinsky reaction. The key message which we wish to convey here it the *modus operandi* which illustrates the basic logical steps that should be utilized on the way to the state-space description of self-organized systems. In passing, we will also see what potential obstacles can be expected along the route.

2   Measurement in Self Organized Systems

2.1   The Problem of Orthogonal Coordinates

The principal component analysis has been first introduced by Pearson as early as in 1901 [4]. Its application consists of three steps: (1) normalization of each data-set, (2) calculation of the covariance matrix and (3) calculation of principal components, i.e. orthogonal coordinates which determine directions of main variance in the data. When principal components are calculated, it may be calculated covariance of each of the original data-sets with each principal components. These covariances are called *loadings*.

Experimentalist's experience is often such, that loadings are naturally coinciding with components of one known process or feature. For example, chemical compounds characterizing one biochemical process have highest loading on the

38   D. Štys et al.

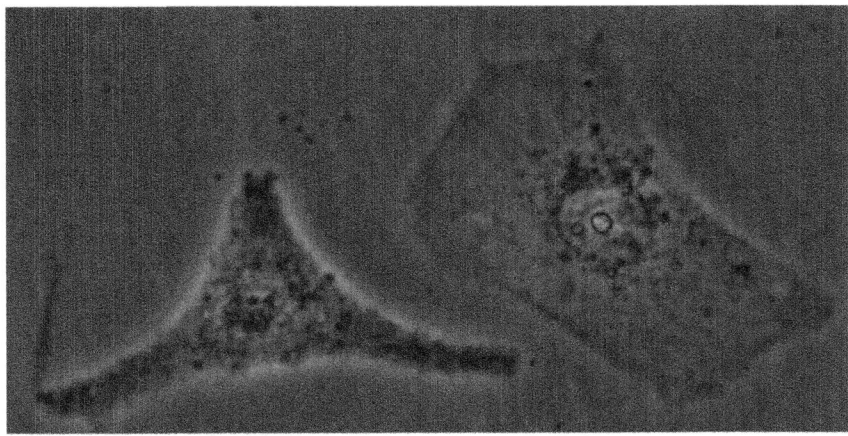

Fig. 2. Two typical structures of living HeLa cell observed by the microscope. The figure comes from Refs. [12,13].

same principle axis. In Ref. [12,13] we have discussed the necessity to character-ize microscopic images of living cells in chemical, as well as, in physical space. The chemical space needs to be characterized with chemical potentials which are in logarithmic relation to activities. In equilibrium thermodynamics, both these variables are related to standard thermodynamical variables of tempera-ture, pressure and volume through systems state function.

It is somehow naturally assumed, that pressure, volume, temperature and number of molecules are orthogonal state coordinates of the system. At least, that is the impression from textbooks on physical chemistry. In fact, the only mathematically correct statement is that state function of the ideal gas is a manifold in the orthogonal state space with coordinates pressure, volume, tem-perature and number of particles. For our discussion it is, however, more valid to discuss the Gibbs free energy. For ideal gas, the state function is a plane — eventually hyperplane, in the space with coordinates G, T and $\wp_i$ (i.e., Gibbs en-ergy, temperature and partial pressures of chemical components in the system). With some manipulations we for the condensed phase case come to widespread formula

$$G \ = \ G_0 \ + \ RT \sum_{i=1}^{n} \ln c_i \,, \tag{1}$$

where c_i are concentrations of system components. G_0 denotes standard Gibbs energy of given system and R the universal gas constant. All equilibrium ther-modynamics is devised in this state space. In chemical thermodynamics, in or-der to obtain more realistic description, activity a_i (or fugacity) replaces the partial pressure of ideal gas component. The correction factor, so-called activity coefficient γ_i, is in function accounting for transformation of coordinates in order to obtain formally the same equation as that for ideal gas.

$$ G \; = \; G_0 \; + \; RT \sum_{i=1}^{n} \ln a_i \; = \; G_0 \; + \; RT \sum_{i=1}^{n} \ln c_i + RT \sum_{i=1}^{n} \ln \gamma_i \, , \qquad (2) $$

Activity coefficient is non-linear function of concentration of all species present in the sample. From physico-chemical point of view, it accounts for all molecular interactions in given system. Generally, manifolds of state functions are accessible only experimentally even for such simple systems as two component mixtures of miscible liquids and their vapours.

Principle components approach [4] looks at the problem from different point of view. The resulting variables, principal components, may be assumed merely as practical descriptors of the system. When the problem is treated from more elementary point of view, the principal components are best approximation to internal orthogonal coordinates in state space appropriate to stochastic (macroscopic) variables of which each has normal distribution. Instead of constructing the state function as manifold in a given coordinate system, we find coordinate system in which the state function is a plane in multidimensional space. When equilibrium physico-chemical system is examined, we may consider principal components best approximations to log-activities as functions of all systems variables.

2.2 Orthogonal Coordinates of in the Phase Space and Generalized Fractal Dimensions

A nice and extensively studied example of self-organizing object is the Lorenz attractor (cf., e.f. Ref. [7]). The beautiful butterfly wings structure is plotted in orthogonal phase-space coordinates which originally had physical meaning of total flux of matter, temperature difference between incoming and outgoing stream and deviation of temperature gradient from linearity. Certainly, these are not parameters which are easily measurable in model Bénard cell, where is to the measurement accessible the diameter (in general shape) of the cell and temperature of the lower and upper plate.

Results of measurements of real objects are to a large degree stochastic. The real system is subjected to external fluctuations and repeated use of for example - mechanical measurement device naturally brings up normal distribution of measured value probability. Digitization of analogue signal has similar effect, however, often leading to distortion of the signal. These are just few of many sources of stochasticity in real measurements.

The quest for proper measurement of self-organizing systems is to describe them by measurable variables which reflect their internal structures. As well as principal components best reflect the axes according to which a given dataset may be best separated. Self-organizing systems, however, are not homogeneous, they have their observable, self-maintained, dynamic structures. Thus the first problem to be addressed, before the multivariate analysis is performed, is to find a method how to characterize the internal structure.

We essentially measure what may be measured by accessible equipment. Or, in more precise expression, the measured system has to be examined only in relation to its phenomenological variables which are the measured sub-set of all possible variables representing system attributes, as was precisely formulated for example by Žampa [8] (see also [12]). The internal, in real system at least within statistical error orthogonal, coordinates, are not accessible to the experimentalist. We proposed earlier [12] to extract the information content of the given data-set using Rényi entropy gain approach. Rényi entropy has a close connection to generalized dimension of the multifractal objects [9,10]. We, at given image resolution yardstick, use Rényi entropies at different parameter, calculated simply from occurrences of different signals, as first attempt to determine measure in the state space.

At this place we discuss self-organizing multifractal object whose stability is based on their character of chaotic attractor. This allows us to build direct analogy with the Lorenz attractor. Each individual multifractal contribution to resulting spatially distributed signal, i.e. image, has its own spectrum of generalized dimensions. There should be as many components as there are dimensions of the phase space.

Each component of the real multifractal object should have its own generalized spectrum. In our practical approach this means that each of the Rényi entropy values at each alpha has its occurrence in the image. We determine the information contribution of each data point to the object by calculation of difference in Rényi entropy of the data-set between the data-set including and excluding the examined object. This is the point information gain (PIG) $PIG_{\alpha,x,y}$ introduced recently in Ref. [12]. For given data-set, i.e. image, the three-dimensional spectrum in coordinates Rényi entropy, the dimensionless α coefficient and occurrence of a given values in the set is a unique characteristic of the image. We are well aware of the fact that the relation of PIG to individual component of the generalized spectrum of individual component of the observed SO is only vague, but we do not have better technical way how to proceed. PIG for given point x,y at given alpha is calculated according to formula

$$\text{PIG}_{\alpha,x,y} = \frac{1}{1-\alpha} \ln \left(\sum_{i=1}^{n} p_{i,x,y}^{\alpha} \right) - \frac{1}{1-\alpha} \ln \left(\sum_{i=1}^{n} p_i^{\alpha} \right). \qquad (3)$$

where $p_{i,x,y}^{\alpha}$ and p_i^{α}) are probabilities of occurrence of given intensity in the image without and with the examined point. In the first approach, calculation of PIG from whole image, distribution of intensities within the object of study is not considered. Then occurrence and entropy represent the same feature. Inclusion of spatial information may be done by defining the reference frame, in our case the cross whose shanks intersect in examined point. We shall show below that the later method allows in the examined case better estimation of orthogonal principle components.

In the next step we calculate integral information as potential state variable. We calculate sum of all PIG values, the point information gain entropy PIE, and point information gain entropy density PIE/points, the sum of all PIG levels at given alpha. Unfortunately it is likely that in non-symmetrical systems PIG, PIE and PIE/ points is dependent on choice of coordinates. This is given by the measurement device, for illustration we may consider that we examine the structure of Lorenz attractor while having only one arbitrary plane from which it is observed.

$$\text{PIE/ points}_\alpha = \sum_{i=1}^{n} \text{PIG}_{\alpha.i}. \tag{4}$$

In this article we, on measurable systems, experimentally examine following hypothesis:

1. The orthogonal coordinates obtained by principle component analysis are practically usefull approximation to the internal orthogonal coordinates of the asymptotically stable system which gives rise to observed self-assembling object.
2. The covariances (loadings) between each calculated PIE/points or PIE values and the given principal component are reasonable and practically usefull approximation to intensities of spectral lines of individual multifractal component, i.e. chaotic attractor - in observed set which gives rise to observed self-assembling object.

3   Results

3.1   Observables in General Projections of a Strange Attractor

An obvious way to determine validity of our approach is to prepare simulated data and expose them to the aforementioned analysis. Recent calculation of fractal dimension of the Lorenz attractor was done by Viswanath [5]. He reported that calculated values are very sensitively dependent on a numerical precision of the implemented calculation. Addition of noise would add additional, and surprising, dimension to such demanding calculation. This we wish to demonstrate on an illustrative example.

In Fig. 3 we show snapshots of the evolution of a toy model of Lorenz attractor [11]. There we show results of simulation in the absence of noise. We further show results of the same simulation with on-negative Gaussian noise applied in all three orthogonal directions. This toy model, in which the numerical integration is performed by the Euler method with a relatively large integration step, shows impressive robustness of the Lorenz attractor system of equations. The influence of Gaussian noise is in a sense puzzling: instead of generating random behaviour of the system it leads to *de-coherence*. When noise is applied in y axis it leads to preference of one wing of the butterfly and when applied in the z axis the system collapses into one point of focus. We do not provide any proof of generality of this observation, however, this result seems to be in agreement with observations of Jalnine *et al.* [15] who predicted collapse of chaotic behaviour in the van der Pol oscillator at increasing noise. Similar results using different approach to noise generation were also reported by Zahri [16]. The model is freely available [11] and the reader is encouraged to explore it. In any case, it illustrates that any simple assumption about emergence of normal distribution may fail in the case of systems based on strange attractors. Obviously, however, the attempt to obtain image of internal orthogonal structure in chaotic attractor including noise would be extremely sensitive to computational performance and we refrained to include it into our analysis. We limit ourselves to experimentally observed SO systems, where the Nature calculates the trajectory for us with best achievable precision. Since we do not have any information on probability density function of the processes contributing to observed phenomena in experimental self-organizing systems, it is necessary to provide some logically feasible distribution. On the basis of the central limit theorem, it is natural to expect that the normal distribution will be the best choice. Under this assumption we use principal component analysis. As a first approximation to the state-space variables we use the Rényi entropy values for individual alpha parameters.

3.2   Measurable Variables in Projections of General Strange Attractor

In Fig.5 we show the basic steps of our analysis. We first calculate information structure of a given image. For each point we calculate the PIG, the Rényi entropy difference at given α calculated from experimentally determined probability density function including and not including the examined point [11,12]. At the moment we have implemented two methods of sampling of the image for the probability density function calculation. First is the obvious approach

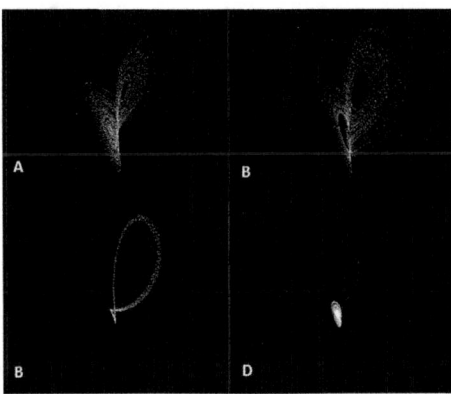

Fig. 3. Examples of behaviour of a toy model of Lorenz attractor. Model was prepared in Netlogo3D [17] program environment. Gaussian random noise was added in three dimensions separately. The noise was added as positive random number with normal distribution. Panel A shows original, noise-free, model. Addition of noise in x axis (panel B) shows expected randomisation of the trajectory. Addition of noise in the y axis, however, leads to formation of narrow stripe of trajectories (panel C) and addition of noise in the z-axis (panel D) leads to collapse of the chaotic attractor structure.

in which the number of points of given intensity is summed and normalised. In the second approach, the image is sampled for each of the points using a cross, whose shanks intersect in the examined point. The probability density function is calculated for the intensity distribution along the shanks. More detailed description of the approach was given in [13,14]. We examined typical images in the course of self-organization of emergent structures of the Belousov–Zhabotinsky reaction.

The question remains whether PIE/points values may be used as coordinates in the phase space. This was examined experimentally. We utilized plain approach to computation of principal components as implemented in Unscrambler software [6]. One of successful attempts is depicted in Fig. 5, where we show decomposition of the system trajectory of the Belousov–Zhabotinsky into series of states which are asymptotically stable under current conditions. The reason why we can not provide one single recipe for the analysis originates comes from the process of observation itself. The geometrical setup affects the course of self-organization. From that follows that proper method of sampling of the image has to be adopted accordingly. In other SO systems, such as living cells or flocks, the effect of geometry may be less pronounces because the geometry is determined by objects themselves. There the example is, however, less instructive because the number of important principal components is typically six or more. In this respect we may only say that such a method was successfully tested on several typical data-sets but its generality has not been proven. In the last step we have examined the contribution of the original Rényi's fractal dimensions to the final ones. In Fig. 5 we depict the contribution of individual components to the whole image for 1st, 2nd and 3rd principal component. The first obvious thing is observation that there is not any qualitative difference between information carried in 1st and 2nd principal component. This may be seen from parallel course of dependency of individual PIE/points covariances when plotted against α. Some differences may be seen in 3rd principal component, where namely the contribution of the blue color deviates significantly. Following the line of arguments given in part in previous section we suggest that the principal components found for each asymptotically stable state is the best approximation to orthogonal coordinates of the chaotic attractor whose reflection are the observed measurable variables. By system state we understand, for example cluster found on the Belousov–Zhabotinsky reaction trajectory. Measurable variables in the case of self organizing systems must include also structure of the system. In the same way as in standard physico-chemical description where a function combining in principle all concentrations in the system makes up the true coordinate - activity, the determined PIE/points values combine into one principal component orthogonal inner coordinate describing macroscopic behavior of the self-organizing system.

Since we have started the analysis from the theoretical assumption of multifractality and calculated first technical approximation to Rényi's entropy, we may suggest that covariances between the PIE/points and principal components

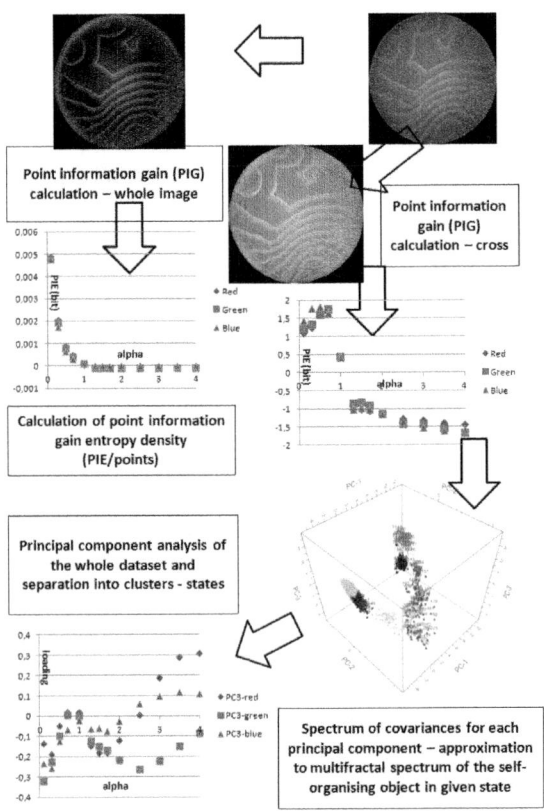

Fig. 4. Scheme of the analysis of experimentally observed self-organized systems. First, information on structure is extracted in the form of point information gain - PIG. Sum of all PIG levels provides point information gain entropy density - PIE/points. The vector of point information entropy densities is used as experimental dataset for the principal component analysis. Data clustering is performed on the data-sets which provides best experimentally achievable separation of the data-set into self-similar clusters - individual asymptotically stable states. For each of the clusters is performed separate analysis which provides, besides other information, spectrum of covariances between each principle component and original PIE/points value for given α parameter. These give, in our opinion, rough impression of how a generalized multifractal spectrum of the underlying attractor in projection to internal orthogonal component may look like.

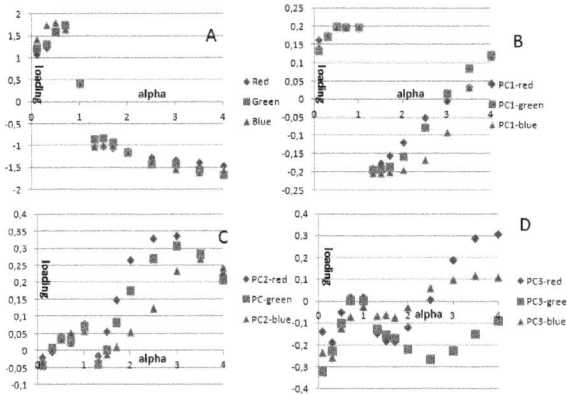

Fig. 5. Various approximations to generalized multifractal dimensions which arise in the course of the analysis. At panel A are given initial PIE/points values for the image 148 of the course of Belousov–Zhabotinsky reaction. Panels B, C and D provide covariances of first, second and third principal component and all utilized PIE/points values for the first cluster (images 56 - 135) which include the image under discussion. We propose that first, second and third principal components represent three independent contributions to the image.

are first technical approximation to multifractal spectrum of asymptotically stable dynamic systems determining systems dynamics. However, both above assumptions need further investigation. We provide all materials regarding the Belousov–Zhabotinsky reaction discussed in this article as well as all necessary software applications. Interested reader is encouraged to examine these primary data.

4   Materials and Methods

4.1   BZ Reaction

We used the kit provided by Dr. Jack Cohen http://drjackcohen.com/BZ01.html. The reaction mixture was composed out of following solutions: A. 25gm sodium bromate dissolved in 335ml water to dissolve to which 10ml of concentrated sulphuric acid (95-98%) was added, B. 10g sodium bromide dissolved in 100ml of water, 10g malonic acid dissolved in 100 ml of water, and D. 1, 10 phenanthroline ferrous complex (Fisons, Loughborough). The reactio was started in following order: 6ml of solution A was added to the glass beaker, then 0.5ml of B, and quickly mixed in 1ml of C. The brown mixture was left to lose bromine (by an open window) until it is pale straw colour or colourless (2-3 minutes if agitated or if in a flat dish). 1ml of the redox indicator D was added, mixed thoroughly and poured into a 9cm glass or plastic petr dish, on a white (illuminated) background. The reaction was followed by colour camera Canon PowerShot G9.

4.2   HeLa Cell Microscopy

HeLa cells (Human Negroid cervix epitheloid carcinoma)were obtained from ECACC - European collection of cell cultures.The cells were grown at low optical density overnight in a 37 °C temperature in a synthetic dropout media with 30% raffinose as the sole carbon source. The nutrient solution for HeLa cells consists of: 86% EMEM, 10% newborn-calf serum, 1% antibiotics and antimycotics, 1% L-glutamine, 1% non essential amino acids, 1% $NaHCO3$ (all components from the PAA company).

In total, 5000 original phase-contrast images of growing cells were taken at 1-min intervals. The image collection was performed at 37 °C by using the Olympus IX 51 with an automated state, integrated incubator, and photographic Camedia C7070 camera. The objective with 20 magnification was used in an image capture.

4.3   Software Packages

For calculation of PIG, PIE and PIE/points values were used software packages developed by the School of Complex Systems. Stable versions of these softwares are available at http://www.expertomica.eu/software.php and are constantly updated. Latest versions are available upon request. For calculation of the Lorenz attractor model was used modification of the model provided to the user community by Salzano in 2005 largely modified by Dalibor Stys [11]. The model is also available at http://www.expertomica.eu/software.php The principal component analysis was performed using Unscrambler software provided by CAMO company [6].

Acknowledgements. This work was partly supported and co-financed by the South Bohemian Research Center of Aquaculture and Biodiversity of Hydrocenoses (CZ.1.05/2.1.00/01.0024)

References

1. http://homepage.tudelft.nl/19j49/
   Matlab_Toolbox_for_Dimensionality_Reduction.html
2. Martens, H.: The Informative Converse paradox: Windows into the unknown. Chemometrics and Intelligent Laboratory Systems, 124–138 (2011)
3. Johnson, R.A., Wichern, D.W.: Applied Multivariate Statistical Analysis. Prentice Hall, London (2007)
4. Pearson, K.: On lines and planes of closest fit to systems of points in space. Philosophical Magazine 2, 559–572 (1901)
5. Viswanath, D.: The fractal property of the Lorenz attractor. J. Atmos. Sci. 20, 130–141 (1963)
6. http://www.camo.com/
7. Lorenz, E.N.: Deterministic nonperiodic flow. Physica D 190, 115–128 (2004)
8. Zampa, P., Arnost, R.: 4th WSEAS Conference (2004)

9. Grassberger, P., Procaccia, P.: Characterization of Strange Attractors. Phys. Rev. Lett. 50, 346–349 (1983)
10. Grassberger, P., Procaccia, P.: Measuring the strangeness of strange attractors. Physica D 9, 189–208 (1983)
11. Stys, D., http://www.expertomica.eu/software.php
12. Stys, D., Vanek, J., Nahlik, T., Urban, J., Cisar, P.: The cell monolayer trajectory from the system state point of view. Mol. BioSyst. 7, 2824–2833 (2011)
13. Stys, D., Urban, J., Vanek, J., Cisar, P.: Analysis of biological time-lapse microscopic experiment from the point of view of the information theory. Micron. 41, 478–483 (2010)
14. Urban, J., Vanek, J., Stys, D.: Preprocessing of microscopy images via Shannon's entropy. Pattern Recognition and Information Processing, 283–187 (2009)
15. Jalnine et al, arXiv:0805.0118v1
16. Zahri, M.: Numerical Solutions of a Stochastic Lorenz Attractor. J. Num. Mat. Stoch. 2, 1–11 (2010)
17. Wilensky, M.: NetLogo. Center for Connected Learning and Computer-Based Modeling, Northwestern University, Evanston, IL (1999), http://ccl.northwestern.edu/netlogo/

Self-Organised Routing for Road Networks

Holger Prothmann[1], Sven Tomforde[2], Johannes Lyda[2], Jürgen Branke[3],
Jörg Hähner[2], Christian Müller-Schloer[2], and Hartmut Schmeck[1]

[1] Karlsruhe Institute of Technology (KIT), Institute AIFB,
76128 Karlsruhe, Germany
{holger.prothmann,hartmut.schmeck}@kit.edu
[2] Leibniz Universität Hannover, Institute of Systems Engineering, Appelstr. 4,
30167 Hannover, Germany
{tomforde,lyda,haehner,cms}@sra.uni-hannover.de
[3] University of Warwick, Warwick Business School, Coventry, CV4 7AL, UK
juergen.branke@wbs.ac.uk

Abstract. Increasing mobility and the resulting rising traffic demands
cause serious problems in urban regions world-wide. Approaches to alle-
viate the negative effects of traffic include an improved control of traffic
lights and the introduction of dynamic route guidance systems that take
current conditions into account. One solution for the former aspect is
Organic Traffic Control (OTC) which provides a self-organised and self-
adaptive system founded on the principles of Organic Computing. Based
on OTC, this paper introduces a novel concept to dynamic route guidance
in urban road networks. Inspired by the well-known protocols Distance
Vector Routing and Link State Routing from the Internet domain, the
major goal of the route guidance mechanism is to increase the network's
robustness with respect to congested or blocked roads. The efficiency of
the developed approach is demonstrated in a simulation-based evaluation
that considers disturbed and undisturbed traffic conditions.

Keywords: dynamic route guidance, traffic signal control, observer/con-
troller architecture.

1 Introduction

Today's urban traffic is characterised by serious congestion problems due to an
increasing demand for mobility. In consequence, the environmental impact of
motorised traffic is becoming a major concern in public debates and scientific
research [9]. Strategies to satisfy the rising demands are manifold and include
a more efficient usage of the existing infrastructure. Promising starting points
are a traffic-responsive control of traffic lights, their self-organised coordination,
and mechanisms for *dynamic route guidance* (DRG).

Considering the dynamic nature of traffic, the distributed location of intersections in urban road networks, and the autonomous behaviour of drivers, the
traffic domain possesses several characteristics that make it an interesting test

F.A. Kuipers and P.E. Heegaard (Eds.): IWSOS 2012, LNCS 7166, pp. 48–59, 2012.
© IFIP International Federation for Information Processing 2012

case for Organic Computing techniques [8]. Earlier work applied the generic observer/controller architecture proposed for Organic Computing to achieve adaptive traffic signal control [7]. The resulting *Organic Traffic Control* (OTC) system is capable of optimising an intersection's signalisation according to the observed traffic flows. The system has been equipped with self-organising coordination mechanisms that allow to establish progressive signal systems (or "green waves") in the network. The resulting signal coordination has been shown to significantly reduce the network-wide number of stops and, in consequence, the fuel consumption and pollution emissions of motorised vehicles.

The existing OTC system tackles traffic in a passive manner by searching for the best signalisation in response to the network's traffic flows. In order to improve the robustness of traffic networks with respect to incidents (like blockages due to accidents or road works), this paper broadens OTC's scope by introducing a self-organising DRG mechanism that actively guides vehicles through the network. The DRG mechanism extends the existing infrastructure with ideas inspired by well-known routing protocols from the data communication domain. Considering the current traffic demand, the routing protocols are modified to determine the best paths through the road network. The recommended routes are then provided to drivers at each intersection of the network. The major goal of the DRG mechanism is to minimise travel times by preventing congestions. Furthermore, a better distribution of traffic streams helps to use the capacity of the road network efficiently.

The remainder of the paper is structured as follows: Section 2 reviews the state of the art in congestion management by discussing various possibilities to dynamically guide drivers through a traffic network. Section 3 revisits the OTC framework that serves as a basis for the proposed self-organising DRG mechanism. The mechanism and its underlying routing protocols are in the focus of Sect. 4. Section 5 discusses the results of a simulation study that evaluates the potential benefits of DRG. Finally, Sect. 6 concludes the paper by giving a summary and an outlook.

2   State of the Art

In today's road networks, GPS-based navigation systems installed in many vehicles guide drivers to their destinations. The systems rely on an internal map of the network which is used by a variant of Dijkstra's algorithm to compute a shortest or fastest route. The route calculation can be based purely on data stored in the map or it can incorporate up-to-date information that is transmitted via the radio's *Traffic Message Channel* (TMC) or a mobile Internet connection. TMC provides digitally coded traffic and travel information via public radio, but covers highways and major roads, only. Data provided via an Internet connection includes urban areas, but its topicality and quality depend largely on the manufacturer-specific penetration rate of a system since the provided data is based on travel times experienced by other drivers.

Other approaches to vehicle routing address the topicality problem with the help of *floating car data*. In [12], Wedde et al. modified their Internet routing protocol *BeeHive* to be suitable for road traffic. In the resulting *BeeJamA* protocol, vehicles are routed from intersection to intersection on a next-hop-basis. The routing is performed by regionally responsible navigation servers that store routing tables for their area and routes to other areas. The routing tables are updated based on information provided by vehicles that are assumed to continuously transmit their position, speed, and destination to the responsible navigation server. Like *BeeJamA*, the DRG mechanism proposed here relies on routing protocols originally developed for the Internet. However, the difference is in the acquisition of data on the current network state. While *BeeJamA* is based on car-to-infrastructure communication, the DRG mechanism proposed here relies on flow and signalisation data that is available at the network's signalised intersections such that no specially equipped vehicles are required.

The integration of route guidance mechanisms into urban traffic control systems has also been investigated in the European COSMOS project [1]. COSMOS developed incident management and rerouting strategies for various adaptive network control systems including MOTION. Rerouting in the MOTION system has been implemented with the help of a macroscopic traffic flow simulator that simulates routing alternatives on-line and recommends optimal routes through the network. Like the mechanism proposed in this paper, routing is based on traffic data available in the control system. However, routing in MOTION relies on centralised simulation and optimisation, while the DRG mechanism proposed here works completely decentralised.

3   Organic Traffic Control

The proposed self-organising DRG mechanism extends the existing OTC system for traffic light control [7]. The OTC system locally optimises an intersection's signalisation at run-time, while neighbouring intersections self-organise to form progressive signal systems in response to the network's traffic flows.

The local optimisation of signal plans at an intersection is based on the observer/controller architecture that has been proposed for Organic Computing [2]. An observer/controller extends a fixed-time or traffic-actuated signal controller – the *System under Observation and Control* (SuOC) – and optimises its signal timings at run-time. The observer monitors the traffic flows of the intersection's turnings and estimates the current vehicular delay for the signalised intersection. The delay is estimated from current flows and signal timings with the help of Webster's approximation formula [11]. Webster's formula is applied for each turning, before the obtained turning delays are combined in a flow-weighted sum to obtain the average vehicular delay for the intersection. Traffic flows and delays are provided to the controller, where they serve as input for a two-levelled learning mechanism. The controller's first level learns to select signal plans on-line in response to the current traffic demand, while the second level performs a model-based off-line optimisation of signal plans. Optimised plans are incorporated in the selection process on the first level.

As several intersections can be located in close vicinity within an urban road network, their coordination is another important aspect. By coordinating the signalisation of neighbouring intersections, progressive signal systems can be established to avoid unnecessary stops. With the help of local communication links, neighbouring intersections exchange data on the measured traffic flows and on their current signalisation. Thereby, traffic streams that largely benefit from coordination can be identified. Within the identified streams, the local signal plans are adapted to obtain a progressive signal system.

The combination of adaptive intersections with a self-organising coordination mechanism provides important technical preconditions for a DRG framework. Turning delays that are derived by an intersection's observer/controller can be reused to estimate travel times of alternative routes under the current traffic demand and signalisation. With the help of the communication infrastructure, turning costs can be distributed in the network. Therefore, the only components missing for a DRG system are a routing protocol that derives and distributes route recommendations and devices (like *Variable Message Signs* (VMS)) that provide the recommended routes to the drivers.

4   Self-Organised Route Guidance

To provide the OTC system with a routing functionality, each observer/controller is extended by a *routing component* (RC). The RCs determine the best routes to prominent destinations in the network. The recommended routes are updated in response to the network's current traffic conditions and are derived with the help of a routing protocol. The decentralised protocol manages the distribution of travel time data in the network and maintains routing tables that contain the currently recommended routes. Drivers are routed on a next-hop-basis from intersection to intersection. The recommended next turn for a prominent destination is announced by VMSs installed at the intersections (or alternatively by means of car-to-infrastructure communication).

The remainder of this section presents two protocols for vehicle routing that are based on the well-known *Distance Vector Routing* (DVR) and *Link State Routing* (LSR) algorithms used in the Internet (see e. g. [10]). The protocols rely on delay estimations provided by the observer/controller components of the network's signalised intersections and reuse their communication network to exchange routing data (see Sect. 3). To handle incidents in the network (like road blockages caused by accidents), an incident detection mechanism is assumed to be available as additional data source. To this end, Klejnowski [6] presented a distributed incident detection technique for urban settings. Before introducing the details and modifications of the DVR and LSR protocols in the remainder of this section, essential differences between routing in communication and road networks are highlighted in the following.

4.1   Differences between Road Traffic Routing and Internet Routing

Although there is a strong relation between road traffic routing and Internet routing, several differences can be observed. In comparison to communication networks, road networks are typically limited in their size (in terms of contained routers or intersections). In addition, the road network of a city is operated by one authority, while communication networks are loosely coupled sub-networks maintained by varying providers. In both cases, the routing decisions are derived by exchanging messages. In data communication, the same channel is used for payload and routing messages, while in traffic routing two separate networks exist which operate at significantly different speeds. Hence, payload (i. e. vehicles) and routing messages do not compete for bandwidth in road networks. However, traffic routing cannot make use of direct channel characteristics, but has to derive travel time estimations from detectors in the road network.

Besides the general structure of the network, nodes themselves are largely different. An intersection in traffic control corresponds to a router in data communication. Typically, a router contains one routing table for all destinations, since no restrictions regarding the relationship between incoming and outgoing links exist. In contrast, intersections can have separate queues for their turnings such that not only the destination of a vehicle, but also the used intersection approach need to be considered. This defines the need of separate routing tables for all approaches of an intersection. In addition, the transmission time within the router can be mostly neglected in data communication, but the occurring delays at intersections are the key characteristic to quantify the link cost (i. e. the travel time).

Furthermore, obvious differences are related to the dissimilarity of data packets and vehicles. Data packets can be dropped in case of overload situations, their ordering can be reorganised according to prioritisations, and some may also be stored for comparably long durations. These options are not available for road networks.

4.2   Distance Vector Routing for Road Networks

The first DRG mechanism adapts the Internet's DVR protocol [10] for the usage in urban road networks. The adapted protocol is responsible for maintaining routing tables for each of the intersection's approaches. It updates them based on routing messages received from neighbouring intersections and communicates changes to its own neighbours.

The mechanism works as follows: Initially, each intersection checks whether it is located in the vicinity of a predefined prominent destination (like the main station, a stadium, or the city hall). If a prominent destination is identified, the RC creates routing table entries for those approaches that can reach the destination via one of the intersection's turnings. The table entries contain the destination, the recommended turning, and the current travel time as cost. The travel time consists of two parts: the distance-induced travel time and the average delay caused by red traffic lights. The former part is approximately static and can be estimated from the length and the speed limits of the connecting road segments (which are assumed to be available at the intersection). The latter value is obtained from the intersection's observer/controller (see Sect. 3). The sum of both values is stored as total travel time for the routing entry.

Once a routing table entry has been created or updated, it is sent as routing message (or *distance vector*) to the corresponding upstream intersection where all routing tables are iteratively updated. To obtain the cost for the received route, the time that is required to reach the sender (i. e. the local turning delay plus the travel time for the connecting road segment) is added to the travel time received with the message. Then, the routing table is checked: If the message's destination is unknown, a new table entry is created from the message. Otherwise, the existing routing entry is checked. If its recommended turning leads to the sender, its travel time is updated. If, on the other hand, the known route recommends a different turning, the travel times of the known and the novel route are compared and the novel route replaces the known one if it is faster. As a result, the routing tables store estimated travel times to each destination.

In literature, the standard DVR protocol from the Internet domain is affected by the *count-to-infinity* problem (see e. g. [10]). The problem occurs in the context of updating the distance vectors. It can be neglected for the modified approach for traffic networks, since (1) data moves much faster than cars, (2) the DRG mechanism is periodically restarted, and (3) traffic situations typically do not change abruptly.

4.3   Link State Routing for Road Networks

A modified LSR protocol serves as DVR alternative. LSR broadcasts estimated travel times for the network's turning movements. Based on these broadcasts, the best routes can be derived.

In a first step, each RC generates a status description for its intersection. For this purpose, *link states* are determined for each path that directly connects a preceding to a succeeding intersection. Every such path includes one turning movement that belongs to the RC's intersection. Besides information about start and end intersection, the link state contains the expected total travel time for the path. Similar to the DVR approach, the total travel time consists of two parts, namely the distance-induced travel time and the delay occurring at the path's signalised turning movement. Again, the former part can be estimated by taking travelled distances and speed limits into account, while the latter part is calculated as the turning's estimated delay using Webster's formula. After determining the link states for all turnings, the RC sends this information to all other RCs in the network using broadcast messages.

After receiving all link state broadcasts, each RC is able to build a graph of the network by combining the link states in a second step. A link state message represents a subgraph describing one intersection and the approaching roads. By connecting these subgraphs according to the defined start and end intersections of each link, each RC obtains a weighted graph that models the road network with the current travel times.

The last step of the process derives minimum-cost routes and stores the route recommendations in the routing tables for each approaching road. Minimum-cost routes may be identified with the help of Dijkstra's algorithm that is applied to the graph generated in the previous step. As result of the calculation, the currently best paths to all destinations are known and can be stored in the routing tables that are associated with the intersection approaches. Like for the DVR protocol, the table entries contain the destination, the recommended next turning, and the estimated travel time to the destination.

5   Evaluation

To evaluate the potential benefits of a DRG system, OTC-controlled intersections with and without their RCs have been compared in a simulation study.

5.1   Experimental Setup

The evaluation has been con-
ducted for a simulated network
that is illustrated in Fig. 1.
The network consists of three
Manhattan-type sub-networks. It
contains 27 signalised intersections (depicted as circles) and
28 prominent destinations (depicted as diamonds). Within each
sub-network, the intersections are
connected by one-laned road segments of 250 m length that provide two additional turning lanes

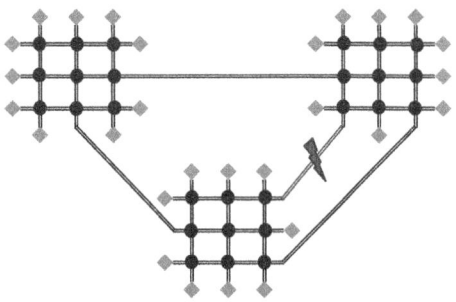

Fig. 1. Network map (incl. incident location)

starting 100 m before an intersection. Regions are connected by two-laned roads.

Signalised intersections are operated by an observer/controller (see Sect. 3) and can provide route recommendations for the prominent destinations. Each destination also serves as origin for traffic entering the network. Two scenarios are investigated:

– In the *regular scenario*, eight vehicles per hour travel from every origin to every destination. In total, 6048 vehicles traverse the network in every hour. Since this demand does not cause significant jams at the network's intersec-tions, the scenario allows to evaluate the impact of DRG under uncongested conditions. It is simulated for a period of three hours.
– In the *incident scenario*, the same amount of traffic traverses the network. However, one of the roads connecting two sub-networks is temporarily blocked due to an incident (see Fig. 1). The blockage affects both directions of the road, occurs after 15 minutes and lasts for 20 minutes within the two hour simulation period. The incident scenario allows to evaluate the benefit of DRG in the presence of disturbances.

As the literature reports a widely varying driver acceptance for VMS-based route recommendations [4, 5], acceptance rates of 0.125 (low), 0.375 (medium),

Table 1. Result summary for the regular scenario

	Ref.	Distance Vector Routing			Link State Routing		
		0.125	0.375	0.75	0.125	0.375	0.75
Travel time [s]	378	342 (9.5 %)	312 (17 5 %)	305 (19.3 %)	348 (7.9 %)	315 (16.7 %)	316 (16.3 %)
Stops [#]	5.14	4.92 (4.3 %)	4.66 (9.3 %)	4.61 (10.3 %)	5.00 (2.7 %)	4.73 (8.0 %)	4.75 (7.6 %)
Fuel [l]	187.6	190.5 (−1.5 %)	182.4 (2.8 %)	178.9 (4.6 %)	192.5 (−2.6 %)	185.0 (1.4 %)	184.9 (1.4 %)
CO [kg]	811.3	784.4 (3.3 %)	721.9 (11.0 %)	707.2 (12.8 %)	797.0 (1.8 %)	731.4 (9.8 %)	731.8 (9.8 %)
HC [kg]	63.9	61.5 (3.8 %)	56.2 (12.1 %)	55.0 (13.9 %)	62.4 (2.3 %)	56.8 (11.1 %)	56.8 (11.1 %)
NO_x [kg]	13.8	13.2 (4.3 %)	12.0 (13.0 %)	11.7 (15.2 %)	13.4 (2.9 %)	12.2 (11.6 %)	12.1 (12.3 %)

and 0.75 (high) have been investigated for both scenarios. The routing protocols have been executed every 150 s and are evaluated with respect to the mean travel time and the mean number of stops per vehicle. These measures indicate how efficiently the road network is utilised and reflect the drivers' comfort. Additionally, fuel consumption and pollution emissions of all vehicles have been investigated in total to estimate the environmental impact of DRG. The evaluation of pollution emissions focusses on Carbon Dioxide (CO_2), Carbon Monoxide (CO), Nitrogen Oxides (NO_x), and un-burnt Hydrocarbons (HC) as these are the main pollutants emitted from petrol and diesel engines. CO, NO_x, and HC are emitted especially during high load and idling periods of petrol and diesel engines (i. e. when vehicles are standing with running engines or when they have to accelerate after a stop). The emission of CO_2 is (for a given type of fuel) directly proportional to the fuel consumption.

All response variables have been evaluated with the help of the microscopic traffic simulator AIMSUN v. 5.1.11 [3]. Fuel consumption rates and pollution emissions have been derived with the help of AIMSUN's environmental models.

5.2 Experimental Results

Simulation results for the regular scenario are summarised in Table 1. The table lists the mean travel time and the mean number of stops per vehicle and gives the total amount of consumed fuel and emitted pollutants for the entire network. To account for stochastic influences in the simulated environment, all table entries are average results taken from five simulation runs.

Table 1 provides data for a reference case without routing and for the proposed DRG mechanisms. In the reference case, simulated vehicles randomly select a route that minimises the distance to their destination. In the DRG case, recommended routes (with respect to current traffic conditions) are provided to the drivers. The different routing protocols and the assumed acceptance rates are listed in separate table columns. Percentage points in brackets specify the relative improvement compared to the reference case. From the results, conclusions can be drawn with respect to the suitability of the routing protocols, the influence of the driver acceptance, and the general advantage of DRG.

Results indicate a benefit of DRG for regular traffic conditions. Independently of the assumed acceptance rate, both routing protocols lead to reduced mean travel times and stops. In consequence, fuel consumption and pollution emissions are reduced in most cases. Only for low acceptance rates, DRG causes a slight increase of the total fuel consumption as the additional fuel used on the faster, but longer recommended routes is not compensated by the relatively small reduction of jams that is achieved by rerouting. This negative effect does not occur for medium and high acceptance rates that exhibit a significantly better performance with respect to all response variables.

Figure 2 visualises mean travel times and stops obtained over the course of the simulation period. The figure shows that DVR and LSR lead to reductions especially during the first two hours. In this period, the observer/controller components at the intersections are still learning. Thus, the signalisation is not yet optimal such that queues can be observed in some road segments. This opens possibilities for DRG. Once all intersections are operated with a near-optimal signal plan, queues are relatively small everywhere in the network. In consequence, the benefit of routing is limited, but still observable.

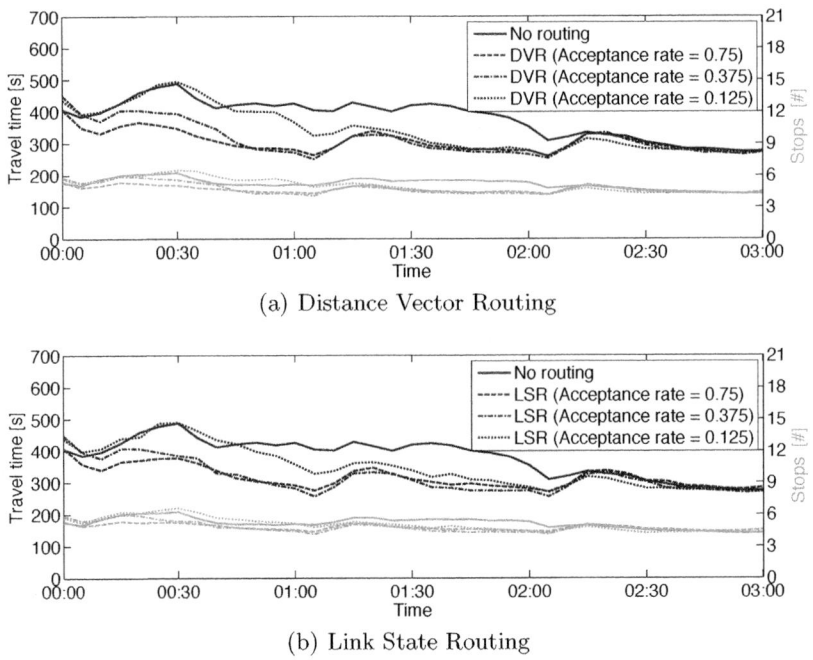

(a) Distance Vector Routing

(b) Link State Routing

Fig. 2. Mean travel times and stops for the regular scenario

In the regular scenario, queues can be reduced by an optimised signalisation alone. This changes for the incident scenario, where traffic jams are caused by blocked roads that disturb the regular traffic flows. Traffic lights can adapt their signalisation to changed flows, but they cannot eliminate growing jams caused by the incident. Here, DRG helps to guide drivers away from blocked roads.

Table 2 summarises the simulation results for the incident scenario. As in the regular case, the implementation of a DRG mechanism is beneficial with respect to the mean travel time and the mean number of stops experienced by individual drivers – the only exception being a marginally increased number of stops observed for DVR at a low acceptance rate. In consequence, the total

Table 2. Result summary for the incident scenario

	Ref.	Distance Vector Routing			Link State Routing		
		0.125	0.375	0.75	0.125	0.375	0.75
Travel time [s]	457	422 (7.7 %)	349 (23.6 %)	333 (27.1 %)	414 (9.4 %)	354 (22.5 %)	340 (25.6 %)
Stops [#]	5.60	5.62 (−0.4 %)	4.97 (11.3 %)	4.83 (13.8 %)	5.56 (0.7 %)	5.05 (9.8 %)	4.94 (11.8 %)
Fuel [l]	204.8	213.7 (−4.3 %)	194.3 (5.1 %)	187.8 (8.3 %)	212.4 (−3.7 %)	197.4 (3.6 %)	193.1 (5.7 %)
CO [kg]	636.4	632.3 (0.6 %)	528.9 (16.9 %)	507.9 (20.2 %)	622.5 (2.2 %)	538.0 (15.5 %)	521.0 (18.1 %)
HC [kg]	51.3	50.4 (1.4 %)	41.8 (18.5 %)	39.9 (22.2 %)	49.6 (3.3 %)	42.4 (17.3 %)	40.9 (20.3 %)
NO_x [kg]	10.9	10.8 (0.9 %)	8.8 (19.3 %)	8.4 (22.9 %)	10.6 (2.8 %)	9.0 (17.9 %)	8.6 (18.9 %)

amount of fuel consumed and pollutants emitted in the network is reduced in most cases. Only at a low acceptance rate, the total fuel usage is increased by rerouting. This can again be attributed to the length of recommended detours and the relatively small contribution to jam reduction in the low acceptance case. Independently of the routing protocol, the system performance improves with respect to all response variables as the acceptance rate increases. Overall, both protocols perform comparably well in the incident scenario.

Figure 3 depicts the mean travel time and the number of stops over the course of the simulation period. The peak that occurs after approximately 35 minutes and lasts for the rest of the first simulated hour indicates that the simulated incident affects both response variables. As the depicted data is gathered when the simulated vehicles have completed their trip, effects of the incidents show up in the figure with some delay after the incidents' occurrence. Furthermore, the disturbances last for some time after the incident has cleared.

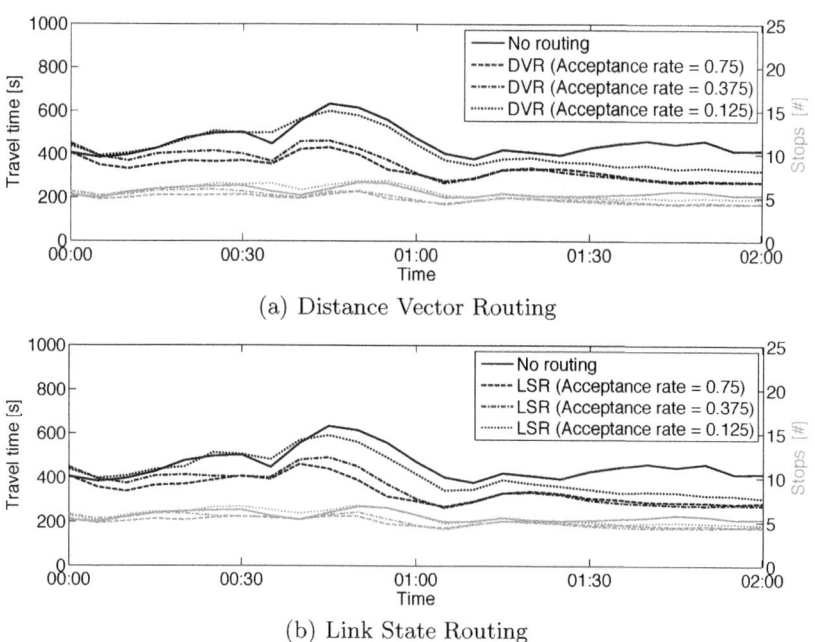

(a) Distance Vector Routing

(b) Link State Routing

Fig. 3. Mean travel times and stops for the incident scenario

When no route recommendations are provided, uninformed drivers try to use temporarily unavailable routes which causes the increase of travel times and stops. With the help of DRG, the incidents' negative effects can be alleviated independently of the applied routing protocol. Informed drivers that comply to the provided route recommendations avoid congested parts of the network. For medium or high acceptance rates, neither travel times nor stops reach the high peak values that can be observed for the reference scenario without DRG. Furthermore, both measures quickly reach their normal levels after an incident. This shows that DRG can improve the robustness of a road network with respect to disturbances like road work or accidents. Reasonably, this positive effect is more pronounced for higher acceptance rates.

6   Conclusion

The paper presented a self-organising approach for dynamic route guidance in urban road networks. An existing framework for traffic signal control has been extended with communicating routing components that derive route recommendations in response to the network's current traffic demand. The routing components are located at signalised intersections, where they estimate turning delays from the locally available traffic and signalisation data. Using a distributed vehicle routing mechanism that is inspired by the *Distance Vector Routing* and *Link State Routing* protocols known from the Internet, drivers are guided from intersection to intersection on a next-hop-basis.

A simulation study has investigated the benefits of the distributed route guidance system. Compared to drivers who randomly pick a shortest route to their destination, a traffic-responsive routing can significantly reduce travel times, stops, and – in consequence – also fuel consumption rates and pollution emissions even for moderate acceptance rates. Although vehicle routing is found beneficial even for regular traffic conditions, its benefits can be observed especially in the presence of blocked road segments. Here, a dynamic routing improves the robustness of the traffic network by guiding drivers around the disturbed areas. The beneficial effects of routing remain (to a limited extent) also for low acceptance rates.

Future work will refine the vehicle routing concept by including hierarchically structured routing tables. By partitioning a large network into regions and introducing intra- and inter-region routing tables, the requirements for the computation, communication, and storage of routing data can be reduced.

Acknowledgment. We gratefully acknowledge the financial support by the German Research Foundation (DFG) within the priority programme 1183 "Organic Computing".

References

1. Bielefeldt, C., Condie, H.: COSMOS – Congestion Management Strategies and Methods in Urban Sites. Final report, The MVA Consultancy (1999)
2. Branke, J., Mnif, M., Müller-Schloer, C., Prothmann, H., Richter, U., Rochner, F., Schmeck, H.: Organic Computing – Addressing complexity by controlled self-organization. In: Margaria, T., Philippou, A., Steffen, B. (eds.) Proc. 2nd Int. Symp. on Leveraging Applications of Formal Methods, Verification and Validation (ISoLA 2006), pp. 200–206 (2006)
3. Casas, J., Ferrer, J.L., Garcia, D., Perarnau, J., Torday, A.: Traffic Simulation with Aimsun. In: Barceló, J. (ed.) Fundamentals of Traffic Simulation, pp. 173–232. Springer, Heidelberg (2010)
4. Emmerink, R.H.M., Nijkamp, P., Rietveld, P., Van Ommeren, J.N.: Variable message signs and radio traffic information: An integrated empirical analysis of drivers' route choice behaviour. Transportation Research Part A: Policy and Practice 30(2), 135–153 (1996)
5. Erke, A., Sagberg, F., Hagman, R.: Effects of route guidance variable message signs (VMS) on driver behaviour. Transportation Research Part F: Traffic Psychology and Behaviour 10(6), 447–457 (2007)
6. Klejnowski, L.: Design and implementation of an algorithm for the distributed detection of disturbances in traffic networks. Master's thesis, Institut für Systems Engineering – System und Rechnerarchitektur, Leibniz Universität Hannover (2008)
7. Prothmann, H., Branke, J., Schmeck, H., Tomforde, S., Rochner, F., Hähner, J., Müller-Schloer, C.: Organic traffic light control for urban road networks. Int. Journal of Autonomous and Adaptive Communications Systems 2(3), 203–225 (2009)
8. Schmeck, H.: Organic Computing – A new vision for distributed embedded systems. In: Proc. 8th IEEE Int. Symp. on Object-Oriented Real-Time Distributed Computing (ISORC 2005), pp. 201–203 (2005)
9. Schrank, D., Lomax, T.: The 2009 Urban Mobility Report. Tech. rep., Texas Transportation Institute (2009)
10. Tanenbaum, A.S.: Computer Networks, 4th edn. Pearson Education (2002)
11. Webster, F.V.: Traffic Signal Settings. Road Research Technical Paper No. 39, UK Road Research Laboratory, Dept. of Scientific and Industrial Research (1958)
12. Wedde, H.F., Lehnhoff, S., et al.: Highly dynamic and adaptive traffic congestion avoidance in real-time inspired by honey bee behavior. In: Holleczek, P., Vogel-Heuser, B. (eds.) Mobilität und Echtzeit – Fachtagung der GI-Fachgruppe Echtzeitsysteme, pp. 21–31. Springer, Heidelberg (2007)

P2P and Cloud:
A Marriage of Convenience for Replica Management[*]

Hanna Kavalionak and Alberto Montresor

University of Trento

Abstract. P2P and cloud computing are two of the latest trend in the Internet arena. They could both be labeled as "large-scale distributed systems", yet their approach is completely different: based on completely decentralized protocols exploiting edge resources the former, focusing on huge data centers the latter. Several Internet startups have quickly reached stardom by exploiting cloud resources, unlike P2P applications which often lack a business model. Recently, companies like Spotify and Wuala have started to explore how the two worlds could be merged, exploiting (free) user resources whenever possible, aiming at reducing the bill to be payed to cloud providers. This mixed model could be applied to several different applications, including backup, storage, streaming, content distribution, online gaming, etc. A generic problem in all these areas is how to guarantee the autonomic regulation of the usage of P2P vs. cloud resources: how to guarantee a minimum level of service when peer resources are not sufficient, and how to exploit as much P2P resources as possible when they are abundant.

1 Introduction

Cloud computing represents an important technology to compete in the world-wide information technology market. According to the U.S. Government's National Institute of Standards and Technology[1], cloud computing is *"a model for enabling ubiquitous, convenient, on-demand network access to a shared pool of configurable computing resources (e.g., networks, servers, storage, applications, and services)"*. Usually cloud services are transparent to the final user and require minimal provider interactions. By giving the illusion of infinite computing resources, and more importantly, by eliminating upfront commitment, small- and medium-sized enterprises can play the same game as web behemoths like Microsoft, Google and Amazon.

The rise of cloud computing has progressively dimmed the interest in another Internet trend of the first decade of this century: the peer-to-peer (P2P) paradigm. P2P systems are composed by nodes (i.e. peers) that play the role of both clients and servers. In this architecture the workload of the application is managed in a distributed fashion without any point of centralization. The lack of centralization provides scalability, while exploitation of user resources reduces the service cost. However, several drawbacks exist, like availability and reliability.

---

[*] This work is supported by the Italian MIUR Project *Autonomous Security*, sponsored by the PRIN 2008 Programme.

[1] http://csrc.nist.gov

F.A. Kuipers and P.E. Heegaard (Eds.): IWSOS 2012, LNCS 7166, pp. 60–71, 2012.

P2P held similar promises with respect to cloud computing, but with relevant differences. While P2P remains a valid solution for cost-free services, the superior QoS capabilities of the cloud makes it more suitable for those who want to create novel web businesses and cannot afford to lose clients due to the best-effort philosophy of P2P.

An interesting "middle ground" combines the benefits of both paradigms, offering highly-available services based on the cloud while lowering the economic cost by exploiting peers whenever possible. Potential applications include storage and backup systems [19,16], content distribution [11,15], music streaming [10], and potentially online gaming and video streaming [2].

In all these systems, a generic problem to be solved is related to the autonomic regulation of resources between the cloud and the set of peers. As an example, consider an online gaming system, where the current state of the game is stored on a collection of machines rented from the cloud. As the number of players increase, it would be possible to transfer part of the game state to the machines of the player themselves, solving the problem with a P2P approach. This would reduce the economical costs of the cloud, yet providing a minimum level of service when peer resources are insufficient or entirely absent.

In order to strike a balance between service reliability and economical costs, we propose a self-regulation mechanism that focuses on replica management in cloud-based, peer-assisted applications. Our target is to maintain a given level of peer replicas in spite of churn, providing a reliable service through the cloud when the number of available peers is too low.

The background needed to understand this work is given in Section 2. Section 3 describes the model of the system and provides a more detailed overview of the research problems associated to this work. The description of the solution is given in Section 4. Section 5 presents the evaluation results, while Section 6 discusses related work. The last section concludes the paper and discusses future work.

2   Background

Unstructured topologies are a promising architectural design for cloud-based, peer-assisted applications. Unstructured P2P networks have indeed a number of attracting properties that suit this kind of applications. Links between nodes are established arbitrarily. To join the network a node just have to acquire a set of links from one of the peers of the network. These important properties are able to reduce the network overload caused by churn.

One of the most notable example of unstructured P2P networks is based on *epidemic* (or *gossip*) protocols [4]. The popularity of these protocols is due to their ability to diffuse information in large scale distributed systems even when large number of nodes are continuously joining and leaving the system. Such networks have proven to be highly scalable, robust and resistant to failures.

One of the fundamental problems of epidemic protocols is how to maintain information about the set of peers participating to the network. Ideally, each node should be aware of all peers in the system, and periodically selects one of them for state exchange.

But in large, dynamic networks this is impossible: the cost of updating all peers whenever a new peer joins or leave is simply too high. To solve this problem, each node maintains a partial view of the network composing the overlay [8,17]. This approach provides each peer with information about a random sample of the entire population of peers in the system, so the network can be represented by a small amount of local information.

As we already mentioned, our work focuses on cloud-based applications. A particular attention is devoted to CLOUDCAST protocol, proposed by Montresor and Abeni [11]. The protocol uses cloud storage to support information diffusion and connectivity in unstructured overlays. The cloud participates in the communication protocol as a normal peer. This heterogeneous architecture adopts CYCLON for peer sampling [17], where the cloud is used to increase reliability of the information diffusion service. The attractive idea from this paper is that the cloud plays the role of static element and the rate of cloud utilization is fixed, no matter what the size of the system is.

In our work, cloud resources are exploited only in the presence of reliability risks. From an operative point of view, our adaptive system is able to switch the behavior between CLOUDCAST and pure CYCLON protocols according to system parameters, reducing the economical effort with respect to previous solutions.

One of the key aspects of autonomic replica management is monitoring of system components. Self-monitoring is vital for self-organized systems, because it allows the system to have a view on its current use and state.

The most popular solutions for autonomous monitoring for unstructured P2P overlays are based on broadcasting algorithms. Updated information about nodes state is broadcast through the system. However, this approach may incur in large network overhead [6].

In this work we adopt the aggregation method proposed by Jelasity et al. [7] to estimate the network size. The protocol continuously provides up-to-date estimation of an aggregate function (in our case the number of nodes in the overlay) computed over all nodes. Time is divided into intervals, so called epochs. At the beginning of each epoch the protocol is restarted. At the end of the epoch each node knows the approximate size the overlay the node had at the beginning of the epoch. The length of a single epoch determines the accuracy of the measure. The convergence rate does not depend on the network size, hence the monitoring cost for each epoch is determined only by the requested accuracy and network churn rate.

3 System Model and Problem Statement

We consider a system composed by a collection of peer nodes that communicate with each other through reliable message exchanges. The system is subject to churn, i.e. node may join and leave at any time. Byzantine failures are not considered in this work, meaning that peers follow their protocol and messages integrity is not at risk. Peers form a random overlay network based on peer sampling service [8].

Beside peer nodes, an highly-available cloud entity is part of the system. The cloud can be joined or removed from the overlay according to the current system state. The cloud can be accessed by other nodes to retrieve and write data, but cannot autonomously initiate the communication. The cloud is pay-per-use: storing and retrieving data with a cloud is connected with monetary costs.

The problem to be solved is the maintenance of an appropriate redundancy level in replicated storage. Consider a system where a collection of data objects are replicated in a collection of peers. A data object is *available* if at least one replica can be accessed at any time, and is *durable* if it can survive failures.

If potentially all peers replicating an object fail or go offline, or are temporary unreachable, the data may become unavailable or even be definitely lost. A common approach to increase durability and availability is to increase the number of replicas of the same data. Nevertheless, the replicas of the same data should be synchronized among each other and retrieved in case of failure. Hence, there is a trade-off: the higher amount of replicas, the higher the network overhead. Furthermore, a peer-to-peer system may have a limited amount of storage available, so increasing the number of replicas of an object can decrease the number of replicas of another one.

An alternative approach is to store a replica of each piece of data in the cloud [19]. This method improves reliability and scalability. Nevertheless, the economical cost grows proportionally to the size of storage space used.

In order to autonomously support the target level of availability and durability and reduce the economical costs we exploit a hybrid approach. The replicas of the same data object are organized into an overlay. To guarantee reliability (durability and availability), the data object must be replicated a predefined number of times. In case P2P resources are not enough to guarantee the given level of reliability, the cloud resources are exploited, potentially for a limited amount of time until P2P resources are sufficient again.

The main goal is thus to guarantee that the overlay maintains a given size, including the cloud in the overlay only when threats of data loss do exist.

In order to realize such system, the following issues have to be addressed:

- To reach the maximum economical effectiveness of the network services, we have to effectively provide mechanisms to automatically tune the amount of cloud resources to be used. The system has to self-regulate the rate of cloud usage according to the level of data reliability (overlay size) and service costs.
- To regulate the cloud usage, each peer in the overlay has to know an updated view of overlay size. Hence, one of the key aspects that has to be considered is the network overhead caused by system monitoring.
- Churn rate may vary due to different reasons such as time of day, day of the week, period of the year, etc. and average lifetime of nodes can be even shorter than one minute. Hence, the size of the overlay can change enormously. Data reliability must be supported even in the presence of one single replica (the replica stored on the cloud), as well as with hundreds of replicas (when the cloud is not required any more and should be removed to reduce monetary costs).

4   The Algorithm

As already mentioned, our goal is to build an unstructured overlay linking the replicas of a data object and maintaining a minimal size in spite of churn. To obtain this, we make use of a number of well-known protocols like CLOUDCAST [11] and CYCLON [17] for overlay maintenance and an aggregation protocol [7] to monitor the overlay size.

We consider the following three thresholds for the overlay size: (1) *redundant* R, (2) *sufficient* S and (3) *critical* C, with $R > S > C$. The computation of the actual threshold values is application-dependent; we provide here only a few suggestions how to select reasonable parameters, and we focus instead on the mechanisms for overlay size regulation, general enough to be applied to a wide range of applications.

- The *critical* threshold C should be the sum of (i) the minimum number of replicas that are enough for successful data recovering and (ii) the amount of nodes that will leave during the cloud backup replication phase. The minimum number of replicas is determined by the chosen redundancy schema, such as erasure coding or data replication.
- The *sufficient* threshold S can be computed as the sum of (i) *critical* threshold and (ii) the amount of nodes that are expected between two successive recovery phases, meaning that these values depends also on the expected churn rate.
- The *redundant* threshold R is an important parameter that allows to optimize the network resources utilization. The *redundant* threshold depends on the overlay membership dynamics (i.e. peers continuously join and leave the network) and on the application type. For example, backup applications are more sensitive to replica redundancy than content delivery applications. In a typical backup application, users save their data in the network and R is determined according to the adopted redundancy schema and overlay membership dynamics.

  On the other hand, in content delivery applications, popular data is replicated in a high (possibly *huge*) amount of nodes of the network. In this case the *redundant* threshold is not relevant, since peers keep pieces of data based on their popularity. Nevertheless, the *critical* and *sufficient* thresholds are important in order to keep data available during the oscillation of network population.

In general, the size of the overlay is expected to stay in the range $[S, R]$. When the current size is larger than R, some peers could be safely removed; when it is smaller than S, some peers (if available) should be added. When it is smaller than C, the system is in a dangerous condition and part of the service should be provided by the cloud.

Figure 1 illustrates the behavior of our system in the presence of churn. The size $N(t)$ oscillates around threshold S, with $k\%$ deviation. To withstand the oscillation, a system recovery process is executed periodically, composed of *monitoring* phase with duration T_m, a recovery phase with duration T_r and *idle* phase of duration T_i. The length of the monitoring phase depends on the particular protocol used to monitor the system, while the idle phase is autonomically regulated to reduce useless monitoring and hence overhead.

Monitoring is based on the work of Jelasity et al. [7]. The authors proposed a gossip protocol to compute the overlay size, that is periodically restarted. The execution of the protocol is divided into epochs, during which a fixed number of gossip rounds are performed. The number of rounds determines the accuracy of the measure. At the end of each epoch, each peer obtains an estimate of the size as measured at the beginning of the epoch.

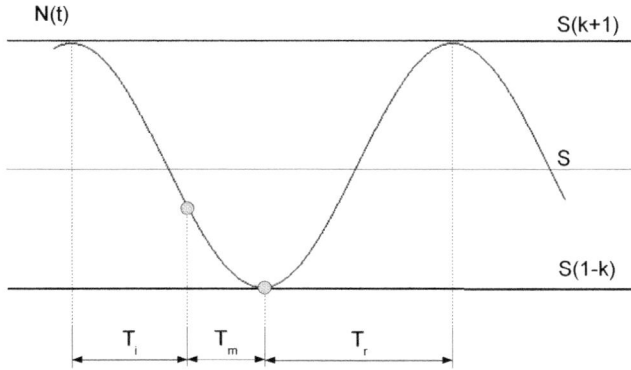

Fig. 1. Overlay size oscillation

We have tuned the aggregation protocol in order to fit our system, with a particular emphasis on reducing network overhead. In our version, each aggregation epoch is composed by monitoring and idle periods. The monitoring period corresponds to the aggregation rounds in the original protocol [7], whereas in idle periods the aggregation protocol is suspended. Our approach allows to reduce network overhead by reducing the amount of useless aggregation rounds.

All peers follow the algorithm in Figure 2. The system repeats a sequence of actions forever. The monitoring phase is started by the call to getSize(), which after T_m time units returns the size of the system at the *beginning* of the monitoring phase. Such value is stored in variable N_b. Given that during the execution of the monitoring phase further nodes may have left the system, the expected size of the system at the *end* of the monitoring phase is computed and stored in variable N_e. Such value is obtained by computing the failure rate f_r as the ratio of the number of nodes lost during the idle phase and the length of such phase.

According to the size of the overlay a peer decides to start the recovering process, delete/create cloud from the overlay or do nothing. If the system size $N_e \geq S$, the peer removes the (now) useless links to the cloud from its neighbor list; in other words, the system autonomously switch the communication protocol from CLOUDCAST to CYCLON. Furthermore, if the system size is larger than threshold R, the excess peers are removed in a decentralized way: each peer independently decide to leave the system with probability $p = (N_e - R)/N_e$, leading to an expected number of peers leaving the system equal to $N_e - R$.

When the number of peers is between critical and sufficient ($C < N_e < S$) a peer invokes function recovery() to promote additional peers to the overlay in a distributed fashion. Each peer in the overlay has to promote only a fraction of additional peers N_a, computed as the ratio of the number of peers needed to obtain the desired $S(1 + k)$ nodes and the current overlay size N_e. The ratio is rounded to the upper bound. Function addNode() adds new peers from the underlying overlay to the set of replicated entities. To bootstrap new peers to the overlay, the peer, currently promoting other peers, copies

```
repeat                                  idleTime(N_e, f_r, T_i)
    N_b ← getSize();                        if N_e ≤ C then
    f_r ← (N_b − N_e)/T_i;                  |  T_i ← T_min;
    N_e ← N_b − T_m · f_r;                  else if f_r > 0 then
    if N_e ≥ S then                         |  T_i ← (S·k)/f_r − T_m;
        removeCloud();                      else if T_i < T_max then
        if N_e > R then                     └  T_i ← T_i + Δ;
        |  p ← (N_e − R)/N_e;               return T_i;
        └  leaveOverlay(p);
                                        recovery()
    else if N_e < S then                    N_a ← ⌈(S(1+k))/N_e − 1⌉;
        if N_e ≤ C then                     for j ← 1 to N_a do
        └  addCloud();                          V ← copyRandom(view);
        recovery()                              if N_e ≤ C then
                                                └  V ← V ∪ {cloud};
    T_i ← idleTime(N_e, f_r, T_i);              addNode(V);
    wait T_i;
```

Fig. 2. Algorithm executed by peers

a set of random descriptors from its own local view copyRandom($view$) and sends them as a V to the new peers. A local view (indicated as $view$) is a small, partial view of the entire network [17].

When the overlay size is below critical ($N_e \leq C$), a peer first adds to the local view a reference to the cloud (function addCloud()) and then starts the recovering process. Moreover, the reference to the cloud is also added to a set of random descriptors V that is sent to the new peers. Hence, in case of a reliability risk, the system autonomously switches the protocol back to the CLOUDCAST (i.e. it includes a reliable cloud node into the overlay).

Concluded this phase, the peer has to compute how much it has to wait before the next monitoring phase starts; this idle time T_i is computed by function idleTime().

If the overlay size is lower than C, the idle time is set to a minimum value T_{min}. As a consequence, the monitoring phase starts in a shorter period of time. Reducing of the idle period allows the system to respond quickly to the overlay changes and keep the overlay size in a safe range. In case the amount of replicas has decreased ($f_r > 0$), the idle time is computed as the ratio of the deviation range $S \cdot k$ and the expected failure rate f_r. The ratio is decreased in T_m time entities needed for the next overlay monitoring phase. If the result T_i is less than T_{min}, the final T_i is set to T_{min}. When either the number of replicas is stable or it has increased ($f_r \leq 0$), the idle period is increased by a fixed amount Δ. The duration of the idle phase is limited by the parameter T_{max}, which is decided by the system administrator.

It is important to note that, in order to increase the amount of overlay peers in the network, a peer has to add them faster then failure rate. In fact, promoting rate of new peers depends on several factors, such as network topology and physical characteristics of the network.

Table 1. Parameters used in the evaluation

Parameter	Value	Meaning
n	0-1024	Total number of peers in the underlying network
R	60	*Redundant* threshold
S	40	*Sufficient* threshold
C	10	*Critical* threshold
k	0.2	deviation from *sufficient* threshold
σ_{cyclon}	10s	Cycle length of CYCLON
$\sigma_{entropy}$	10s	Cycle length of anti-entropy
$\sigma_{failure}$	1s	Cycle length of peers failure
c	20	View size
g	5	Cyclon message size
Δ	1s	Incremental addition step for the idle period T_i

5   Evaluation

The repairing protocol has been evaluated through an extensive number of simulations based on event-based model of PEERSIM [12]. Due to limited space, we present here a brief evaluation of the behavior of the system, leaving an extended description to future work. Unless explicitly stated, the parameters that are used in the current evaluation are shown in Table 1. Such parameters are partially inherited from the CYCLON [17], CLOUDCAST [11] and Aggregation protocols [7]. The choice of the thresholds parameters in Table 1 is motivated by graphical representation; in reality, the protocol is able to support both smaller and bigger sizes of the overlay.

The first aspect that has to be evaluated is protocol scalability: is our protocol able to support overlay availability in spite of network size oscillation? Figure 3 and Figure 4 shows the behavior of the overlay when network size oscillates between 0 and 1024 peers in a one-day period. The average peer life time is around 3 hours.

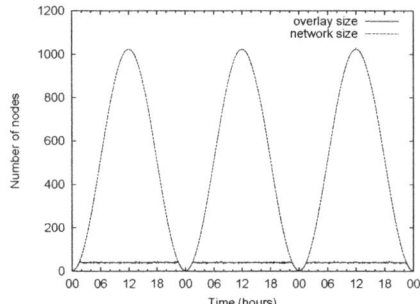

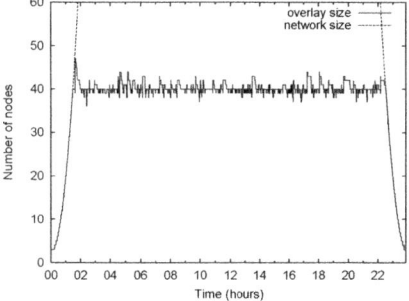

Fig. 3. Overlay size in case of network oscillates daily between 0 and 1024 peers. The single experiment lasted 3 days.

Fig. 4. Overlay size in case of network oscillates daily between 0 and 1024 peers. Zooming over a single day.

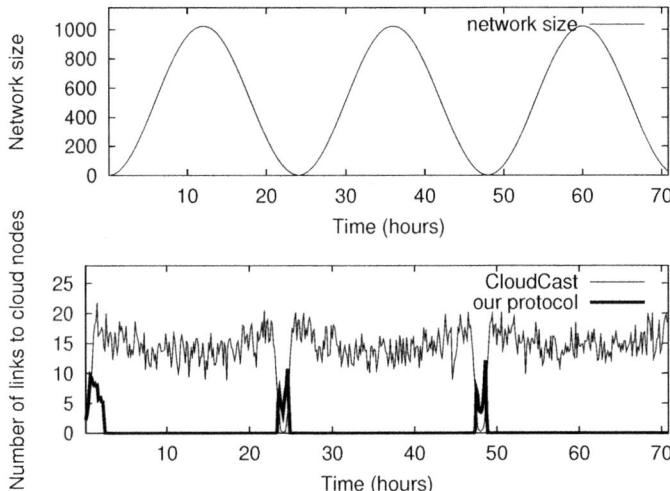

Fig. 5. Cloud in-degree for CLOUDCAST and our protocols in case of network oscillates daily between 0 and 1024 peers.

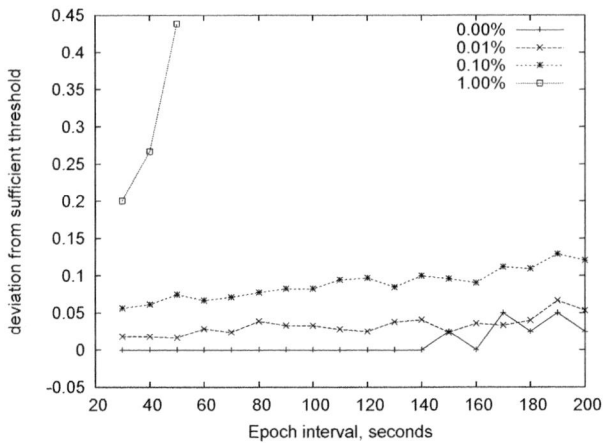

Fig. 6. Deviation of the overlay size from the *sufficient* threshold under different levels of churn with variable epoch intervals.

When the network size grows, the overlay peers promote new peers to the overlay and its size growths until reaching the sufficient threshold. Then the protocol supports the sufficient overlay size with k deviation. When all peers leave the network, the overlay size falls to zero and contains only the cloud peer. This behavior repeats periodically.

Figure 5 shows the amount of existing links to the cloud in the overlay. The top figure shows the network size oscillation and the bottom one represents the cloud in-degree for CLOUDCAST (thin line) and our protocol (thick line). As you can see, compared with CLOUDCAST, our protocol significantly reduces the utilization of cloud resources without implications on the overlay reliability. When the network size is enough to provide the *sufficient* overlay size, references to the cloud are removed; otherwise the overlay peers include cloud node into the overlay.

For system monitoring we adopt the work of Jelasity et al. [7]. Figure 6 shows the ability of the protocol to support the target overlay size under the various network churn rates and epoch intervals. The X-axis represents the duration of the epoch intervals. The Y-axis represents the deviation of the overlay size from the target *sufficient* threshold. A churn rate $p\%$ for the overlay means that at each second, any peer may abruptly leave the overlay with probability $p\%$. In this evaluation we do not consider that peers can rejoin the overlay.

As we can see from Figure 6, the lines correspond to 0%, 0.1% and 0.01% shows approximately the same behavior. These results proof that the system successfully supports the target deviation ($k = 0.2$). However, with the churn rate 1% the system is not able to support the overlay with an epoch interval larger then 30 seconds. Nevertheless, it must be noted that the churn rate of 1.0% corresponds to an average lifetime of less then 2 minutes.

6 Related Work

One of the most important limitations in P2P-based storage systems is connected with the difficulty to guarantee data reliability. P2P nodes leave or fail without notification and stored data can be temporarily or permanently unavailable.

A popular way to increase data durability and availability is to apply redundancy schemas, such as replication [9,3], erasure coding [18] or a combination of them [13]. However, too many replicas may harm performance and induce excessive overhead. Therefore it becomes a relevant issue to keep the number of replica around a proper threshold, in order to have predictable information on load and performance of the system.

Two main approaches exist for replica control: (1) proactive and (2) reactive. Proactive approach creates replicas at some fixed rate [14,13], whereas in reactive approach a new replica is created each time an existing replica fails [9,3].

The work of Kim [9] proposes a reactive replication approach that uses lifetime of nodes to select where to replicate data. When the data durability is below a threshold the node does replicas in nodes whose lifetime is long enough to assure durability. Conversely, Chun et al.[3] proposes a reactive algorithm, where data durability is provided by selecting a suitable amount of replicas. The algorithm responds to any detected failure and creates a new replica if the amount of replicas is less then the predefined minimum.

Nevertheless, these approaches are not taking in account the redundancy of replicas in the network. In fact, increasing the number of replicas affects bandwidth and storage consumption [1].

The problem of replica control is a relevant problem and one of the most complete work in this area is presented by Duminuco et al. [5]. To guarantee that data is never lost the authors propose an approach that combines optimized proactive with pure reactive replication. The approach is able to dynamically measure churn fluctuations and adapt the repair rate accordingly.

Nevertheless, Duminuco et al. consider the estimation of recovering rate, whereas in our work the attention is devoted for the periods of recovery restarting and network overloads caused by system monitoring. Moreover, Duminuco et al in their work do not consider the distributed mechanism of regulation the overlay population.

Finally, recent works in the field do not consider the oscillation of the network size and the situation when P2P resources are not enough or not available. To solve these problems we use the combination of both communication models: P2P and cloud Computing. Moreover, we propose a mechanism that automatically includes or excludes the cloud into the overlay, by adapting the system to the network state.

7   Conclusions

In this paper, we have considered a hybrid peer-to-peer/cloud network, where a replicated service is provided on top of a mixed peer-to-peer and cloud system. We have designed a protocol that is able to self-regulate the amount of cloud (pay-per-use) resources when peer resources (free) are not enough. Future work include considering a full-fledged solution for one of the application areas listed in the introduction, such as distributed storage, online gaming or video streaming.

References

1. Blake, C., Rodrigues, R.: High availability, scalable storage, dynamic peer networks: pick two. In: Proc. of the 9th Conf. on Hot Topics in Operating Systems, Lihue, Hawaii. USENIX (2003)
2. Carlini, E., Coppola, M., Ricci, L.: Integration of P2P and clouds to support massively multiuser virtual environments. In: Proc. of the 9th Workshop on Network and Systems Support for Games (NetGames 2010), Taipei, Taiwan. IEEE Press (2010)
3. Chun, B.-G., Dabek, F., Haeberlen, A., Sit, E., Weatherspoon, H., Frans Kaashoek, M., Kubiatowicz, J., Morris, R.: Efficient replica maintenance for distributed storage systems. In: Proc. of the 3rd Conf. on Networked Systems Design & Implementation (NSDI 2006), San Jose, CA. USENIX (2006)
4. Demers, A.J., Greene, D.H., Hauser, C., Irish, W., Larson, J., Shenker, S., Sturgis, H.E., Swinehart, D.C., Terry, D.B.: Epidemic algorithms for replicated database maintenance. In: Proc. of the 6th ACM Symposium on Principles of Distributed Computing Systems (PODC 1987), pp. 1–12 (1987)
5. Duminuco, A., Biersack, E., En-najjary, T.: Proactive replication in distributed storage systems using machine availability estimation. In: Conference on Emerging Network Experiment and Technology (2007)
6. Idreos, S., Koubarakis, M., Tryfonopoulos, C.: P2P-diet: an extensible P2P service that unifies ad-hoc and continuous querying in super-peer networks. In: Proc. of the 2004 Int. Conf. on Management of Data, SIGMOD 2004, Paris, France, pp. 933–934. ACM (2004)
7. Jelasity, M., Montresor, A., Babaoglu, O.: Gossip-based aggregation in large dynamic networks. ACM Trans. Comput. Syst. 23(1), 219–252 (2005)

8. Jelasity, M., Voulgaris, S., Guerraoui, R., Kermarrec, A.-M., van Steen, M.: Gossip-based peer sampling. ACM Trans. Comput. Syst., 25 (August 2007)
9. Kim, K.: Lifetime-aware replication for data durability in P2P storage network. IEICE Trans. on Communications E91-B, 4020–4023 (2008)
10. Kreitz, G., Niemelä, F.: Spotify – large scale, low latency, P2P music-on-demand streaming. In: 10th IEEE Int. Conf. on Peer-to-Peer Computing (P2P 2010), pp. 1–10, Delft, The Netherlands (August 2010)
11. Montresor, A., Abeni, L.: Cloudy weather for P2P, with a chance of gossip. In: Proc. of the 11th IEEE P2P Conference on Peer-to-Peer Computing (P2P 2011), pp. 250–259. IEEE (August 2011)
12. Montresor, A., Jelasity, M.: PeerSim: A scalable P2P simulator. In: Proc. of the 9th Int. Conference on Peer-to-Peer (P2P 2009), Seattle, WA, pp. 99–100 (September 2009)
13. Pamies-Juarez, L., Garcia-Lopez, P.: Maintaining data reliability without availability in P2P storage systems. In: Proc. of the ACM Symp. on Applied Computing, SAC 2010, Sierre, Switzerland, pp. 684–688. ACM (2010)
14. Sit, E., Haeberlen, A., Dabek, F., Chun, B.-G., Weatherspoon, H., Morris, R., Kaashoek, M.F., Kubiatowicz, J.: Proactive replication for data durability. In: 5th Int. Workshop on Peer-to-Peer Systems, IPTPS 2006 (2006)
15. Sweha, R., Ishakian, V., Bestavros, A.: Angels in the cloud – A peer-assisted bulk-synchronous content distribution service. Technical Report BUCS-TR-2010-024, CS Department, Boston University (August 2010)
16. Toka, L., Dell'Amico, M., Michiardi, P.: Online data backup: A peer-assisted approach. In: IEEE 10th Int. Conf. on Peer-to-Peer Computing, pp. 1–10 (2010)
17. Voulgaris, S., Gavidia, D., van Steen, M.: CYCLON: Inexpensive membership management for unstructured P2P overlays. J. Network Syst. Manage. 13(2) (2005)
18. Williams, C., Huibonhoa, P., Holliday, J., Hospodor, A., Schwarz, T.: Redundancy management for P2P storage. In: Proc. of the Seventh IEEE Int. Symp. on Cluster Computing and the Grid, CCGRID 2007, pp. 15–22. IEEE Computer Society, Washington, DC (2007)
19. Yang, Z., Zhao, B.Y., Xing, Y., Ding, S., Xiao, F., Dai, Y.: Amazingstore: available, low-cost online storage service using cloudlets. In: Proc. of the 9th Int. Workshop on Peer-to-Peer Systems, IPTPS 2010, San Jose, CA. USENIX (2010)

Self-Organizing Spatio-temporal Pattern Formation in Two-Dimensional Daisyworld

Dharani Punithan and R.I. (Bob) McKay

Structural Complexity Laboratory, Seoul National University, South Korea
{punithan.dharani,rimsnucse}@gmail.com

Abstract. Watson and Lovelock's daisyworld model [1] was devised to demonstrate how the biota of a world could stabilise it, driving it to a temperature regime that favoured survival of the biota. The subsequent studies have focused on the behaviour of daisyworld in various fields. This study looks at the emergent patterns that arise in 2D daisyworlds at different parameter settings, demonstrating that a wide range of patterns can be observed. Selecting from an immense range of tested parameter settings, we present the emergence of complex patterns, Turing-like structures, cyclic patterns, random patterns and uniform dispersed patterns, corresponding to different kinds of possible worlds. The emergence of such complex behaviours from a simple, abstract model serve to illuminate the complex mosaic of patterns that we observe in real-world biosystems.

1   Introduction

Self-organization is a property of processes, in which patterns emerge at global scale due to the short-range local interactions of the system, with no external intervention [2]. A wide range of pattern formation processes have been observed in fields from chemistry (Turing's reaction-diffusion system [3], Belousov-Zhabotinsky reaction [4]) and fluid dynamics (Bénard convection cells [5]) to biology (patterns on animals' coats [6]).

In this paper, we investigate the emergence of self-organized, spatio-temporal vegetation patterns in two-dimensional (2D) Daisyworld [1]. First of all, we show that the surface temperature of planet is self-regulated around the mean 295.5K by the daisyworld dynamics of our model as in previous 2D daisyworld models [7, 8]. Then we observe the self-organizing properties of daisyworld, and qualitatively analyse pattern formation in daisyworld with Laplacian diffusion. We also quantify the emergence of vegetation patterns using a statistical measure, Moran's I. Pattern formation in daisyworld has been previously touched on in both 1D [9] and 2D [10] cellular automaton models. However our daisyworld model is extended to a coupled 2D chaotic system; we investigate daisyworld in diffusively coupled logistic systems in a 2D tordoidal regular lattice with chaotic local dynamics, a setting which has not been studied previously.

In our daisyworld model, depending on model parameter values we observe a variety of behaviours, such as 1) complex patterns (stationary/dynamic), 2) smoothly dynamic spatio-temporal patterns (Turing-like structures), 3) cyclic patterns (periodic spirals), 4) random patterns (stationary/dynamic) and 5) dispersed patterns (uniform).

F.A. Kuipers and P.E. Heegaard (Eds.): IWSOS 2012, LNCS 7166, pp. 72–83, 2012.

2 Background

2.1 Daisyworld

Daisyworld was proposed by Watson and Lovelock [1] as a means to study planetary ecodynamics. This simple self adaptive system has one environmental variable (temperature) and two types of life (black and white daisies). The colour of the daisies influences the albedo (i.e. the reflectivity for sunlight) of the planet, and thereby influences the temperature of the planet. Though the daisies do not interact with each other directly, they do interact via the planetary temperature (environment); the behaviour of one type of daisies modifies the temperature, which affects the behaviour of the other, and vice versa. In general, increased growth of a particular type of daisy alters the temperature in a direction which is unfavourable for itself, but suitable to the other type. When black daisies start spreading more, the local temperature rises. Conversely, when white daisies spread, the temperature drops. Due to the contrary behaviours of the species, the temperature is self-regulated and daisies persist on the planet.

2.2 Turing Instabilities

Turing instability of a homogeneous steady state in a two-species reaction-diffusion system provides theoretical mechanisms for pattern formation. Generally diffusion is perceived as an homogenising process. By contrast, Turing, in his remarkable paper(1952 [3]), demonstrated that a simple system of coupled reaction-diffusion equations could give rise to spatial patterns in chemical concentrations, through a process of chemical instability. The chemicals (morphogens or species), characterised as activator and inhibitor, react and diffuse throughout the tissue, and due to this physico-chemical process patterns such as spots and stripes can form in an animal's coat. These spatial patterns emerge from a homogeneous equilibrium state, due to symmetry-breaking via diffusion. This diffusion-driven instability forms the basis of pattern formation in physics, chemistry and biology. A wide spectrum of complex self organized patterns such as stationary, waves, spirals or turbulence emerge from reaction-diffusion systems.

2.3 Moran's I

Moran's I [11] is a measure of spatial autocorrelation, and is often used to analyse spatial patterns. The specific values -1, 0, +1 indicate spatially dispersed, random, and clustered patterns respectively. It is defined by equation 1:

$$I = \frac{C}{\sum_i \sum_j w_{ij}} \frac{\sum_i \sum_j w_{ij}(P_i - \bar{P})(P_j - \bar{P})}{\sum_i (P_i - \bar{P})^2} \tag{1}$$

where C is the number of spatial units(cells) indexed by i and j, P is the size of the population, $\bar{P}$ is the mean population and w_{ij} is a spatial weight matrix. In our work, $w_{ij} = 1$ if i and j are horizontal or vertical neighbours, and $w_{ij} = 0$ otherwise.

3 Self-Organization and Pattern Formation in Daisyworld

3.1 Positive-Negative Feedback

Positive and negative feedback are the two basic modes of interactions among components of self-organizing systems [2]. Self-enhancing positive feedback coupled with antagonistic negative feedback leads to the formation of striking patterns in nature (ripples in sand dunes, schooling of fish, flocking of birds, coat of animals). A stationary state becomes unstable due to amplification, and this provokes pattern in the system. Since amplification purely in one direction leads to destruction, negative feedback takes control of the system and plays a critical role in inhibiting amplification and shaping the process and pattern, for example in colonial nesting of male bluegill sunfish [2].

In daisyworld, an initial growth of black daisies, due to positive feedback, increases the temperature of system and thus reinforces the change in the same direction. Hence it has the potential to explode the system. But in daisyworld, at a higher temperature, white daisies bloom and keep the positive feedback under control. In the same manner, the growth of white daisies, due to positive feedback, decreases the temperature in the same direction. But at a colder temperature, black daisies bloom and inhibit the system behaviour. The initial growth of daisies promotes new changes in the system state but on the other hand, increased growth counteracts the changes. The combined effect of positive and negative feedback couplings result in a negative feedback loop, which stabilises the physiological processes and leads to homeostasis [12]. Due to this behaviour, we can see that daisyworld is a self-organizing system.

3.2 Activator-Inhibitor Principle

Turing pattern formation in activator-inhibitor systems forms the paradigm of self-organization [13], and provides a theoretical explanation of animal coat patterns [6]. The activator and inhibitor species destabilize the homogeneous state of the system due to diffusion, and lead to the spontaneous emergence of spatial patterns.

We can view daisyworld as a Turing reaction-diffusion system of two morphogens: black and white daisies. In cooler temperatures, black daisies absorb the available sun light and amplify their growth, functioning as activators (autocatalysis) because they activate their own production. But further increase in black daisies increases the surface temperature to the level at which cooler white daisies grow well, thereby acting as an inhibitor (self-suppression) because they inhibit their own production. This process is reversed for white daisies. Depending on the temperature, black and white daisies act both as inhibitor and as activator. Here, the diffusion of the two species, coupled by nonlinear local rules, result in the generation of patterns and spatial self-organization due to symmetry-breaking instability.

4 Model

The interacting components (cells) of the system are arranged on a toroidal 2D square ($N \times N$) regular lattice. Each cell is designated as a habitat with a maximum carrying capacity of 10,000 individuals. All sites in the lattice are randomly initialised with a population size in $[0, 100]$ for both black and white daisies. The temperature is initialised randomly depending on the overlap between the optimal temperatures of black and white daisies (refer Table 2). The interaction rules among these system components are based only on local information (microscopic level) and not on the global emergence (macroscopic level). The black and white daisies, along with the temperature, diffuse around the world via their neighbourhood. The daisyworld model has two parts: the local component describing the effects on growth of daisies and change in temperature local to the particular cell; and the interaction component describing the migration of black and white daisies and diffusion of heat to their neighbours.

Table 1. Daisyworld Parameter Settings (last two rows are variable parameters, others are fixed)

Parameter	Value(s)	Parameter	Value(s)
Fixed			
Number of cells	100×100	Bare ground Albedo (A_g)	0.5
Heat Capacity (C)	2500	Albedo of black daisies (A_b)	0.25
Diffusion constant (D_T)	500	Albedo of white daisies (A_w)	0.75
Stefan-Boltzmann constant $(\sigma_B)\ E^{-8} W m^{-2} K^{-4}$	5.67	Mean growth temperature	295.5 K
Luminosity (L)	1	Solar Insolation (S) $W m^{-2}$	864.65
Variable			
Diffusion rate of black daisies	$[0, 1]$	Diffusion rate of white daisies	$[0, 1]$
Noise Level	$[0, 10]$	Natural Rate of Increase	$[1, 4]$

Table 2. Optimal temperature overlap

Overlap(%)	Opt. temp. (K) black daisies	Opt. temp. (K) white daisies	Initial values (K)
0	278	313	$[270, 320]$
10	284.5	306.5	$[280, 310]$

Local Dynamics: The population size of each unit is computed based on the logistic equation with carrying capacity, refer equation 2:

$$P_{(t+1)} = P_t(1 + r[1 - \frac{P_t}{K}]) \tag{2}$$

where r is the bifurcation parameter (i.e. intrinsic capacity for increase). The parameter r provides positive feedback and the component $[1 - \frac{P_t}{K}]$ provides the negative feedback, with their combined effect regulating the population. Though the regulatory mechanism is built in, chaos emerges from the system [14]. The temperature change at each location is based on the energy balance equation [15] (refer equation 3):

$$\sigma_B T^4 = SL(1 - A) \tag{3}$$

where σ_B is the Stefan-Boltzmann constant, T is the temperature, S is the solar constant, L is the luminosity and A is the albedo.

Global Emergence: The behavioural rules in the system specify that black daisies grow better in colder temperatures, while white daisies grow better in hotter temperatures, both diffusing to their nearest neighbours. These simple components at the local level generate complicated dynamics at the global level due to the collective behaviour of the whole system, without the influence of any external forces. The interaction pattern is determined by the neighbourhood topology and the diffusion term. In our model, we use von Neumann neighbourhoods, consisting of a central cell and its four orthogonal neighbours (top, down, left and right, without diagonal interactions), and a Laplacian diffusion model for both species and temperature. The key ingredients leading to global emergence are the local nonlinear interactions among the components, physiological and behavioural rules, and physical constraints. The inhibition of self-growth can also arise from the physical constraints, due to the self-limiting factor of carrying capacity.

Albedo: The albedo of the planet can be computed by equation 4:

$$A = A_b \alpha_b + A_w \alpha_w + A_g \alpha_g \tag{4}$$

where $\alpha_b, \alpha_w, \alpha_g (= 1 - \alpha_w - \alpha_b) \in [0, 1]$ are the relative areas occupied by black, white daisies and bare ground, A_b is the albedo of ground covered by black daisies, A_w that of ground covered by white daisies and A_g that of bare ground. We assume that $A_w > A_g > A_b$, with corresponding values of $0.75, 0.5, 0.25$.

Temperature: The local temperature update is defined by diffusion, heat radiation, solar absorption and Gaussian noise, as in equation 5:

$$C \cdot \delta T_{(i,j,t)} = D_T \bigtriangledown^2 T_{(i,j,t)} - \sigma_B T^4_{(i,j,t)} + SL(1 - A_{(i,j,t)}) + C\epsilon_{(i,j,t)} \tag{5}$$

where $C = 2500$ is the heat capacity, $\delta T_{i,j,t} = T_{i,j,t+1} - T_{i,j,t}$ is the change in temperature at (discrete) time t, $1 \leq (i, j) \leq N$ indicates a 2D lattice point, N is the lattice length, $D_T = 500$ is the diffusion constant, $\bigtriangledown^2 T$ is the Laplacian operator, σ_B is the Stefan-Boltzmann constant, S is the solar constant, L is the luminosity, A is the albedo and ϵ represents Gaussian white noise (with mean zero and standard deviation 1.0) multiplied by the noise level.

Growth: The growth rate of daisies is defined as a parabolic function as in equation 6:

$$\beta(T) = 1 - [\frac{(T_{opt} - T)^2}{17.5^2}] \tag{6}$$

where T is the local temperature, and T_{opt} is the optimal temperature of the species, which depends on their colour. The optimal temperature for black daisies is lower than for white. In general, growth rates of daisies depend on local temperature, which in turn depends on the albedo and thus on the proportions of the daisy species and bare ground.

Population Size: The local population update is computed based on Laplacian diffusion and local population growth, governed by the logistic equation with carrying capacity, as in equation 7:

$$\delta P_{(i,j,t)} = D \bigtriangledown^2 P_{(i,j,t)} + r P_{(i,j,t)} [\beta(T) - \frac{P_{(i,j,t)}}{K}] \tag{7}$$

where at location (i,j) and time t, $P_{i,j,t}$ is the population size, D is the fraction of the population being dispersed to its neighbours, r is the natural rate of increase, $\beta(T)$ is feedback coefficient, and K is the carrying capacity. Deterministic chaos can also be induced by diffusion and hence our system doesn't require high intrinsic growth rates.

4.1   Assumptions and Limitations

The model parameters are represented in Table 1. They reflect the general daisyworld literature and parameter values [1,7]. The last two rows of Table 1 are variable parameters, while the rest are fixed parameters. The range of temperatures in which daisies can survive is the viability constraint of our model, and is based on the overlap between the optimal temperatures of black and white daisies as in Table 2. We define the optimal temperature of black daisies within the range $[278, 295.5]$K and of white in the range $[313, 295.5]$K so as to have an overlap in the range $[0, 100]\%$. For this work, we use a 10% overlap for most of experiments, with a 0% overlap for one. As we have a huge parameter space, we have restricted our experiments to an increase rate of $r = 1$. We used the same diffusion constant for both black and white daisies in all our experiments. We limited our focus to studying the effects of species diffusion (which influences the species distribution) and noise (which influences the temperature) in detail.

5   Results

The results presented here are samples from over 15,000 experiments with different variable parameter settings, including species increase rates in the range $[1, 4]$, species dispersion rates in $\{0, 0.001, 0.01, 0.1, 0.2, 0.5, 1\}$, overlap in $\{0, 5, 10, 25, 50, 100\}\%$, noise level in $\{0, 0.001, 0.005, 0.01, 0.05, 0.1, 0.2, 0.5, 1, 2.5, 5, 7.5, 10\}$, and temperature diffusion in $\{0, 0.05, 0.1, 0.2, 0.3, 0.5, 1\}$. We examined the results for the emergence of interesting patterns both qualitatively (by capturing snapshots of each experiment and inspecting visually) and quantitatively (using the value of Moran's I). Our preliminary investigation found interesting behaviours in the 0% and 10% overlap settings, and we present these here. Our future aim is to develop suitable metrics to automatically select interesting cases from the parameter space.

We present the results by showing the state of the Daisyworld at different epochs. Since the behaviour is different in each case, the epochs chosen differ in each case. In these visualisations, a location is shown as black if the population of black daisies at a particular location is larger than that of white, and vice versa. The variable parameter values used in each case are detailed in the caption of the figure.

As it is impossible to show all spatial snapshots over 5000 epochs, we have plotted the temporal dynamics of the global temperature, which characterises the nature of the daisyworld. We note that in all cases, the temperature self-regulates around 295.5K, confirming the persistence of the daisies as an emergent property of daisyworld. We have also plotted the value of Moran's I for both black and white daisies.

A wide spectrum of patterns emerge from our model, depending on the model parameter values. We have classified them into 1) complex patterns (stationary/dynamic), 2) smoothly dynamic spatio-temporal patterns (Turing-like structures), 3) cyclic patterns (periodic spirals), 4) random patterns (stationary/dynamic) and 5) dispersed patterns (uniform). We were able to find examples of most behaviours with an overlap of 10%; however for Turing-like structures, we used an overlap of 0%.

5.1   Self-organized Complex Pattern

A maze-like self-organizing pattern (refer Fig. 1) emerge from our model with a diffusion constant 0.001 for both species and a noise level of 0.001. Though the initial states look chaotic, the system finally converged to a stationary maze-like structure. The global spatial behaviour is plotted in Figure 2. The temperature initially fluctuates, but then converges, exhibiting a stationary pattern formation. Initially Moran's I is very high, indicating the formation of very large clusters, but it finally drops below 0, indicating a uniformly dispersed pattern. In between, the fluctuations show complex dynamics.

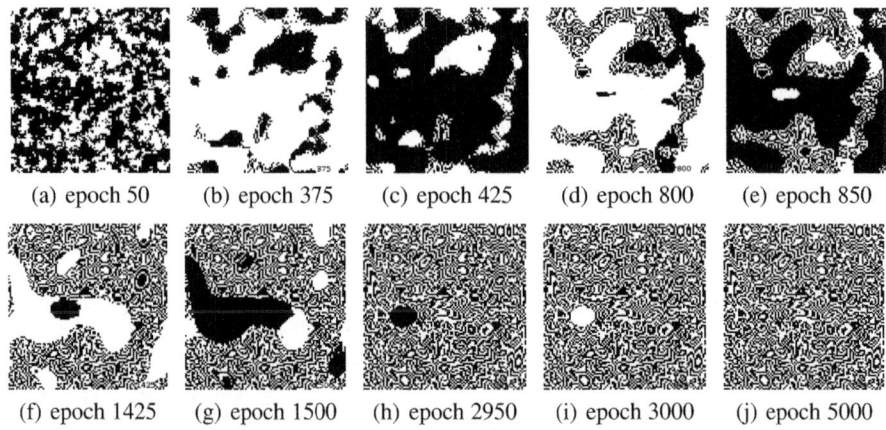

(a) epoch 50	(b) epoch 375	(c) epoch 425	(d) epoch 800	(e) epoch 850
(f) epoch 1425	(g) epoch 1500	(h) epoch 2950	(i) epoch 3000	(j) epoch 5000

Fig. 1. Maze like organization for $D = 0.001$ and noise level$= 0.001$ in 2D 100×100 Lattice

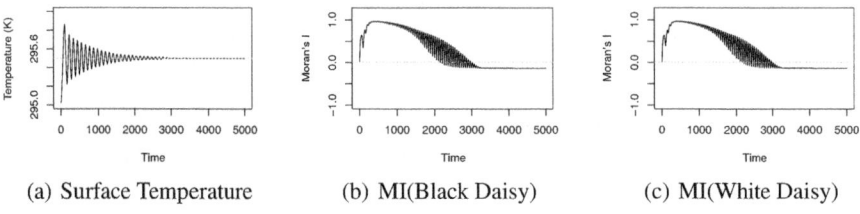

(a) Surface Temperature   (b) MI(Black Daisy)   (c) MI(White Daisy)

Fig. 2. Global dynamics of maze like self-organized pattern formation

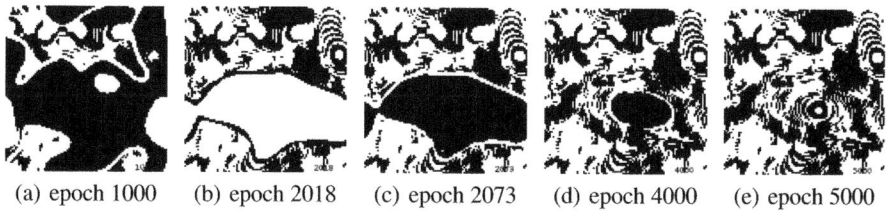

(a) epoch 1000   (b) epoch 2018   (c) epoch 2073   (d) epoch 4000   (e) epoch 5000

Fig. 3. Modern art complex pattern for $D = 0.1$ and noise level$= 0.001$

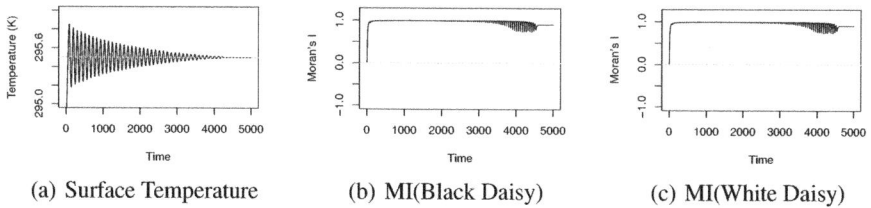

(a) Surface Temperature   (b) MI(Black Daisy)   (c) MI(White Daisy)

Fig. 4. Global dynamics of modern art complex pattern

Raising the diffusion constant to a still relatively low 0.1 for both species, and with a noise level of 0.001, we see the emergence of a 'modern art' complex pattern (Figure 3); Figure 4 shows its global dynamics. Moran's I stays very high throughout the run till the final epoch – and visually we can see clusters in some locations in that final epoch.

5.2 Turing-Like Structures

Turing-like patterns emerge with a diffusion constant of 0.5 for both species, a noise level of 0.2 and an overlap of 0% – refer to Figure 5. The corresponding global dynamics are plotted in Figure 6. The patterns which emerge are smooth and slowly changing, looking like an animal's coat (e.g. Holstein cow). The fluctuations in temperature in sub-figure (a) of Fig. 6 show the dynamicity of pattern formation. The high Moran's I throughout the run (except the initial transitions) corresponds to the emergence of large clusters in the patterns (see subfigures (b) and (c) of Figure 6).

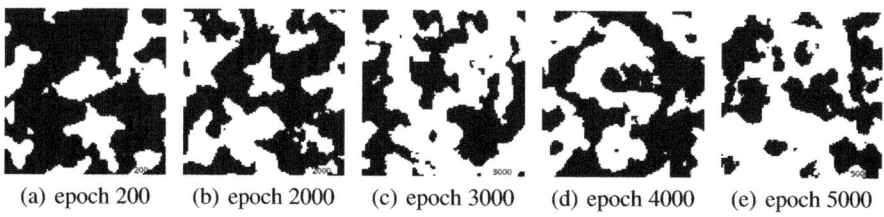

(a) epoch 200   (b) epoch 2000   (c) epoch 3000   (d) epoch 4000   (e) epoch 5000

Fig. 5. Turing-like structures for $D = 0.5$, noise level$= 0.2$ and overlap0%

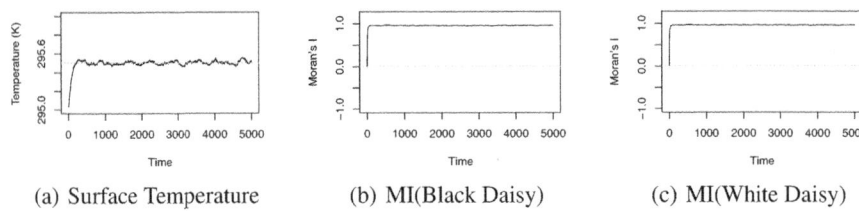

(a) Surface Temperature   (b) MI(Black Daisy)   (c) MI(White Daisy)

Fig. 6. Global dynamics of Turing-like structures

5.3 Periodic Spirals

Periodic spirals can be seen in the behaviour when we use a diffusion constant of 0.2 for both species and a noise level of 0.05 (Figure 7). The corresponding global dynamics are plotted in Figure 8. The cycles in temperature seen in sub-figure (a) confirm the periodicity in pattern formation. Except for the initial states, Moran's I is high throughout the run, confirming the formation of large clusters (sub-figures (b) and (c)).

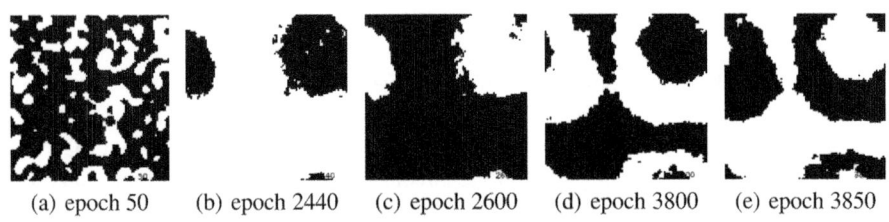

(a) epoch 50   (b) epoch 2440   (c) epoch 2600   (d) epoch 3800   (e) epoch 3850

Fig. 7. Periodic spirals for $D = 0.2$ and noise level$= 0.05$

5.4 Random Pattern

Random patterns evolve from the system for a noise level of 0.01 and with no species dispersion – see Figure 9 and the corresponding global dynamics in Figure 10. The near-zero value of Moran's I in sub-figures (b) and (c) of Fig. 10 confirms the formation of random patterns.

5.5 Dispersed Pattern

Uniformly dispersed patterns emerge at a of noise level 0.5 with no species diffusion (Figure 11); the corresponding global dynamics are depicted in Figure 12. Moran's I in sub-figures (b) and (c) of Figure 12 lie below zero, indicating dispersed patterns. Interestingly, we do see the emergence of a hint of a large scale grid in this setting, presumably resulting from the effects of temperature diffusion.

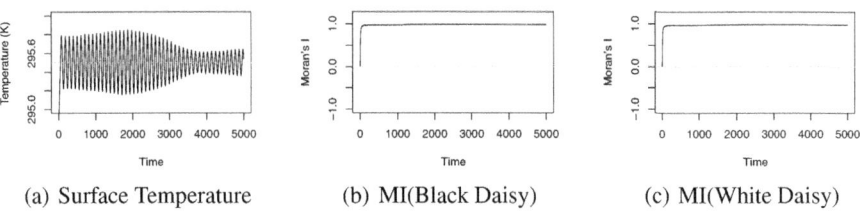

(a) Surface Temperature   (b) MI(Black Daisy)   (c) MI(White Daisy)

Fig. 8. Global dynamics of periodic spirals

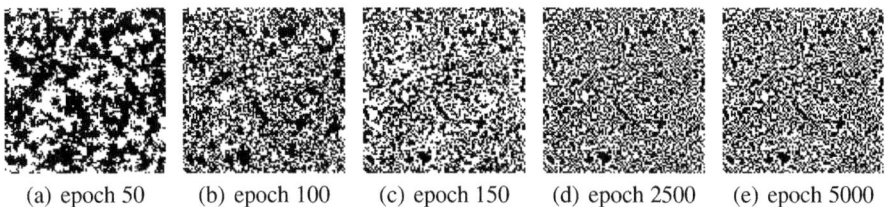

(a) epoch 50   (b) epoch 100   (c) epoch 150   (d) epoch 2500   (e) epoch 5000

Fig. 9. Stationary random for $D = 0$ and noise level 0.01

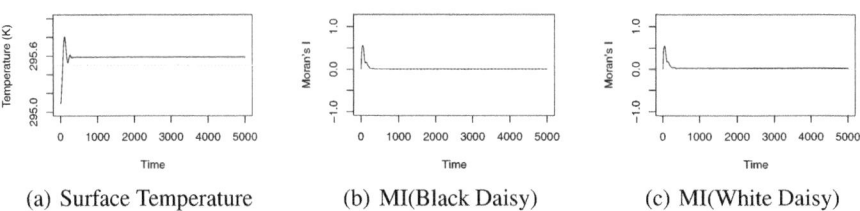

(a) Surface Temperature   (b) MI(Black Daisy)   (c) MI(White Daisy)

Fig. 10. Global dynamics of random pattern

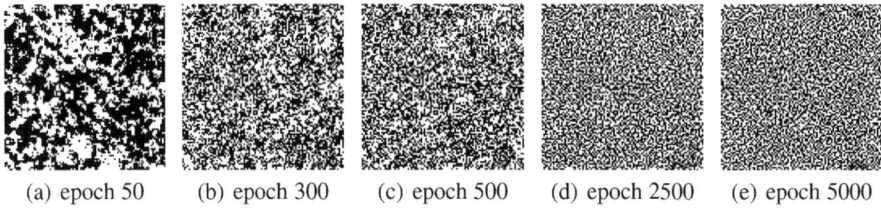

(a) epoch 50   (b) epoch 300   (c) epoch 500   (d) epoch 2500   (e) epoch 5000

Fig. 11. Uniformly dispersed pattern for $D = 0$ and noise level 0.5

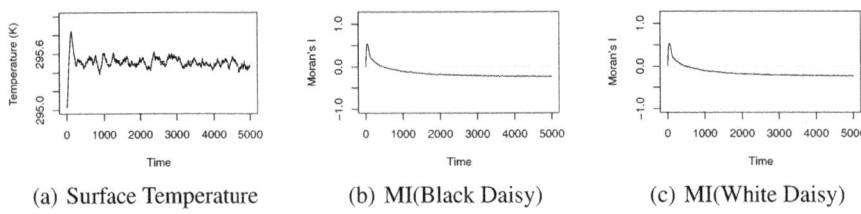

(a) Surface Temperature   (b) MI(Black Daisy)   (c) MI(White Daisy)

Fig. 12. Global dynamics of uniformly dispersed pattern

6   Conclusions

6.1   Summary

The results illustrate that the underlying dynamics of daisyworld are not restricted to stationary patterns but extend to periodic and chaotic behaviour. We see stability as well as complexity emerge. The figures clearly show the spatial patterns and temporal behaviour of daisyworld. Moran's I quantifies the pattern formation accurately, demonstrating its reliability in spatially quantifying a wide spectrum of patterns.

The localised interaction of species (black and white daisies) in our daisyworld model generate an array of fascinating spatio-temporal patterns, due to endogenous self-organization resulting from "symmetry-breaking instability". In our nonlinear reaction-diffusion system, symmetry-breaking is induced by species dispersion, temperature diffusion and noise. A low species dispersion rate combined with noise leads to complex self-organized patterns (Figures 1 and 3). As species dispersion and noise increase, periodic (Figure 7) and dynamic chaotically evolving patterns (Figure 5) emerge. In the absence of species dispersion, random (Figure 9) and dispersed (Figure 11) patterns emerge. We thus see the constructive roles of spatial diffusion, nonlinear local rules and noisy fluctuations in influencing the physiological, behavioural rules and physical constraints and thereby self-organizing the daisyworld.

6.2   Future Work

Since our parameter space is huge, it is practically infeasible to categorise all domains of pattern formation in daisyworld by qualitative visualisation – some automated criterion is required. We used Moran's I, but it assists only with spatial analysis, omitting the temporal analysis. In order to to fully analyse both parameter and pattern formation space, we need objective statistical measures which can uniquely classify the patterns in both space and time. We hope that further work will help us to understand the intrinsic connection between spatio-temporal dynamics and pattern formation in daisyworld.

Acknowledgements. This research was supported by the Basic Science Research Program of the National Research Foundation of Korea (NRF) funded by the Ministry of Education, Science and Technology (Project No. 2011-0004338), and the BK21-IT program of MEST. The ICT at Seoul National University provided research facilities for the study. We would like to thank Ilun Science and Technology Foundation for their generous support to Dharani Punithan.

References

1. Watson, A.J., Lovelock, J.E.: Biological homeostasis of the global environment: The parable of daisyworld. Tellus B 35(4), 284–289 (1983)
2. Camazine, S., Deneubourg, J.-L., Franks, N.R., Sneyd, J., Theraula, G., Bonabeau, E.: Self-organization in biological systems, 2nd edn. Princeton University Press (2003)
3. Turing, A.M.: The chemical basis of morphogenesis. Philosophical Transactions of the Royal Society of London. Series B, Biological Sciences 237(641), 37–72 (1952)
4. Belousov, B.P.: A periodic reaction and its mechanism. Sbornik Referatov po Radiatsonno Meditsine (Medgiz, Moscow), 145–147 (1958) (in Russian)
5. Bénard, H.: Les tourbillons cellulaires dans une nappe liquide. Revue générale des Sciences pures et appliquées 11, 1261–1271 and 1309–1328 (1900)
6. Murray, J.D.: Mathematical Biology II: Spatial models and biomedical applications, 3rd edn., vol. 2. Springer, Heidelberg (2008)
7. von Bloh, W., Block, A., Schellnhuber, H.J.: Self-stabilization of the biosphere under global change: A tutorial geophysiological approach. Tellus B 49(3), 249–262 (1997)
8. Ackland, G.J., Clark, M.A., Lenton, T.M.: Catastrophic desert formation in daisyworld. Journal of Theoretical Biology 223(1), 39–44 (2003)
9. Adams, B., Carr, J., Lenton, T.M., White, A.: One-dimensional daisyworld: Spatial interactions and pattern formation. Journal of Theoretical Biology 223(4), 505–513 (2003)
10. Ackland, G.J., Wood, A.J.: Emergent patterns in space and time from daisyworld: a simple evolving coupled biosphere-climate model. Philosophical Transactions of the Royal Society A 13: Mathematical, Physical and Engineering Sciences 368(1910), 161–179 (2010)
11. Moran, P.A.P.: Notes on continuous stochastic phenomena. Biometrika 37(1/2), 17–23 (1950)
12. Kump, L.R., Kasting, J.F., Crane, R.G.: The Earth System, 3rd edn. Prentice Hall (2010)
13. Nakao, H., Mikhailov, A.S.: Turing patterns in network-organized activator-inhibitor systems. Nature Physics 6(7), 544–550 (2010)
14. May, R.M.: Simple mathematical models with very complicated dynamics. Nature 261(5560), 459–467 (1976)
15. McGuffie, K., Henderson-Sellers, A.: A Climate Modelling Primer, 3rd edn. Wiley, Chichester (2005)

Heuristic Resource Search in a Self-Organised Distributed Multi Agent System

Muntasir Al-Asfoor, Brendan Neville, and Maria Fasli

School of Computer Science and Electronic Engineering, University of Essex,
Wivenhoe Park, Colchester, UK, CO4 3SQ
{mjalas,bneville,mfasli}@essex.ac.uk

Abstract. The work presented in this paper has addressed the issue
of resource sharing in dynamic heterogeneous Multi Agent Systems as
a search problem. When performing a random search, this might lead
to traverse the whole network and increase the failure ratio. This paper has introduced heuristic directed search based on the usage of an
approximate matching mechanism to overcome this problem. Our implementation of search algorithms differs from traditional algorithms
by using semantically guided technique for resource search as well as a
dynamically re-organisable network of agents. The experimental results
have shown that using directed search techniques is better than random
search in terms of number of hops to find the match. Furthermore, network re-organisation has improved the system performance by directing
the search based on resources information, especially when high accuracy
is required.

Keywords: Multi Agent Systems, Resource Sharing, Self-Organisation.

1 Introduction

The rapid development in the scientific research and business requirements for
computational and data resources have increased the need to manage the process
of describing and discovering these resources in an efficient way. Furthermore,
any suggested solution should take into consideration the key characteristics of
pervasive networking which are: heterogeneity, scalability and dynamic behavior.

Resource sharing in a distributed environment has been addressed by many
researchers in the industry and academia. Many frameworks have been developed
to facilitate the process of sharing a geographically distributed resource. In [13] a
dynamic search algorithm in a peer-to-peer system has been proposed, it meant
to overcome the traditional flooding and random search algorithms like searching the whole network aggressively and the long search time respectively. The
proposed algorithm is a modified version of the traditional algorithms (random
walk and flooding) by maintaining a probability function which provides some
knowledge to the peer from the previous states so it could use this knowledge to
predict the best move.

F.A. Kuipers and P.E. Heegaard (Eds.): IWSOS 2012, LNCS 7166, pp. 84–89, 2012.

A dynamic network topology adaptation technique has been proposed by [3] with the aim of improving the search performance of the flooding algorithm by creating semantic communities based on query traffic. In contrast with our work, the authors have focused more on changing the topology based on statistical heuristics by collecting nodes with similar interests together in small communities. They assumed that the file request could be fulfilled by nodes within the same community.

Inspired by the idea of reducing the matching accuracy to discover the resources quickly, we have adapted the conventional random search algorithm by introducing a directed search technique using the available resources description. The novelty we have introduced is to guide the search through the agents' network based on how semantically close the agents are to the requester in terms of resources. The rest of the paper has been organised as follows: section two has been devoted to the resource search scenario. System evaluation and experimental results are presented in section three. Finally, conclusions and suggestions for future development are discussed in section four.

2   Resource Sharing Scenario

In order to support our investigation into the micro-macro link between the selected search parameters, and their global consequences with respect to our system performance metrics, we have chosen to run multiple simulations of our resource discovery scenario across the available state space as defined by our parameters. A number of agent-based simulation tools exist such as Mason [7], NetLogo [12] and Swarm [6] in addition to specific agent languages [9] and platforms e.g. Jason [2]. From this myriad of tools we have selected to use PreSage [8], a Java based rapid prototyping and simulation tool developed to support the animation of agent societies. Similarly to the platforms and tools mentioned above, PreSage allows us to script a series of simulation runs with varying parameters, and record results from a global perspective. The system has been designed as a distributed multi agent system where each agent could be a provider or requester. Each agent holds resources which are available to share or tasks they need to be executed. Each agent has a unique identifier and knows about a subset of the agents' population and uses them to maintain its own contact database. The contacts database contains a list of agents to which the agent can forward messages. The network of agents starts with one agent and then agents will be added to the network while the simulation is running. When a new agent joins the system it will be given a random peer from the set of peers which are already active in the system and then the agent will start to forward and receive messages. The receiver agents will use the senders' contact information to update their contact databases. The agents are interacting using messages of the form:
$Message = < To, From, Resource, Accuracy, TTL >$

where: To: is the sender's ID; $From$: is the potential receiver's ID; $Resource$: is the required resources; $Accuracy$: is the required matching accuracy; TTL: is the maximum time to live (number of hops).

For the purpose of simulation, the contacts database size have been chosen to be finite and each agent can contact k agents directly giving that all the agents are connected in the network layer. Limiting the contacts database size helps to avoid the situation in which each agent is connected directly to all the agents in the network which makes each agent a messages sink; this situation will overload the network and exploit its resources. The simulation's time has been divided into time cycles where a new agent is created each time cycle(t). In our experiments, 300 agents have been created during the first $300t$ of the total $600t$ simulation time.

2.1   Resource Description and Matchmaking

In a distributed computing environment, many standard methods have been developed to describe resources and tasks extensively. For instance, WSDL (Web Service Description Language) [11], OWL (Web Ontology Language) [10] and RDF(Resource Description Framework) [4]. For the purpose of simulation and since the focus of this paper is on improving the network traffic performance, resources and tasks have been described as a three element vector in the form $resource < r_0, r_1, r_2 >$. Where, the vector $resource$ represents the semantic information which then will be used to match resources vs tasks. The resource elements r_i are assigned a random value in the range [0-4] from a uniform distribution which creates 5^3 possible different values. Resource vs. request matchmaking takes place using the Manhattan [5] distance measure between resources and tasks vectors as shown in Equation 1.

$$Dis(R^A, R^B) = \sum_{i=0}^{2} \left| r_i^A - r_i^B \right| \tag{1}$$

where: Dis is the Manhattan distance between resource R^A and resource R^B.

2.2   Resource Search

The resource discovery problem has been considered as a search problem with the aim of finding the required resource across the network. As the network could scale too largely, the aim is to direct the search using the available resource information and by applying the matchmaking technique. Directing the search would decrease the number of hops required to discover the resources and increase the chances of doing that before the request message times out.

As the size of the contacts database is finite, two different methods have been used to deal with the process of adding a new contact to the database. If a new agent has joined and there is a free space in the contacts' database, then it just adds the new contact to the database. Otherwise, if there is no free space in the contacts database, then depending on which adding method has been selected, the agent would either remove the oldest contact and add the new one (First In First Out) in its place or, remove the agent with the lowest similarity and add the new comer (if the new comer is semantically closer than the one to be removed). Technically, the first method (First in First out) helps to maintain the network connectivity of the newcomer and gives it a chance to communicate with other agents. In contrast, the second method (join based on similarity) helps to re-organise the network by directing the search based on the type of resources.

Upon receiving a resource search message (rsm) the receiver agent has to check if the sender is unknown (not available in the contact database), then it adds the sender to the contact database, otherwise, it just updates its record. Furthermore, the receiver will check if the required resource is available locally within the required matching error using Equation 1. Accordingly, if the matching error is less than the maximum error, then the receiver will send a "resource found" message to the initiator and stop. Otherwise, the receiver will have to check if the rsm has timed out or no by checking the time to live parameter of the rsm. After that, if the rsm has timed out, then the search fails otherwise the receiver will have to select the next peer to forward the message to it after updating the path and decreasing the time to live. If there are no more agents to send the rsm to, the agent must send the rsm back to the previous agent in order to enable it to select another branch to send the rsm to.

The next agent(peer) selection is based on two different methods. The first method is to select an agent randomly from the contact database to forward the rsm to excluding the agents which have been already visited before (member of the path). The second method is to select an agent which holds the resource with the smallest matching error with the required resources (given that the error is greater than the maximum allowable error) to forward the rsm to excluding the ones who have already been visited. The modifications we have made during the course of this work are related to the network's connectivity where it changes dynamically when new peers join the system or an existing peer sends an rsm. Even with the conventional random search algorithm, the network's connectivity changes using first In First Out (FIFO) technique to avoid the aging problem we have discussed before. For further details and the Algorithms pseudo code see [1].

3   System Evaluation and Experimental Results

Three experiments have been designed to measure the fidelity of the three configurations with which each agent decides which agent to forward the resource search message to. These configurations are: Config-1: Random Search with First In First Out (FIFO) contacts; Config-2: Directed Search with First In First Out (FIFO) contacts; Config-3: Directed Search with Self-Organisation. For simulation purposes four parameters have been set as follows: Number of Agents = 300; maximum number of contacts for each agent = 10; maximum number of hops each resource search message can do (ttl) = 100; required accuracy of matching between [0,12]. The aim of these experiments is to evaluate the effects of each configuration on the system performance which has been measured by the following parameters: System Mean Average Error (MAE); Average Hop Count to find the resource successfully; Percentage of Failures : the ratio of failed requests to the total number of requests.

In the experiment we ran with required accuracy of 0, it appeared that the matching error between the acceptable and the achieved accuracy was the same for all configurations. The reason for that is simply because there is no matching error less than 0. Furthermore, as the acceptable error increased, the system MAE increased, with directed search approaches doing better than non-directed search as shown in Figure 1(a). However, the average number of hops to find the resource successfully has increased sharply when the acceptable matching error decreased because with smaller acceptable error the message has to navigate more agents to find the required resource. Directed Search with Self-organisation has shown a rapid decrease in the Average Hop Count to find the resource successfully in comparison with Random Search and a noticeable decrease in comparison with Directed Search especially with small acceptable error values as shown in Figure 1(b).

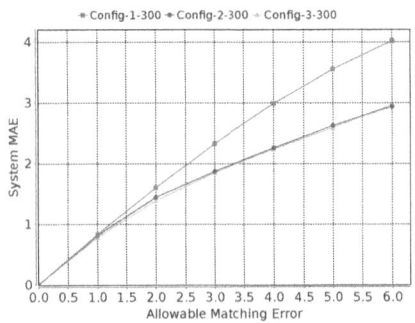 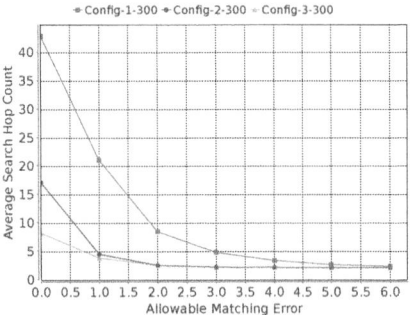

(a) A comparison between the allowable matching error the found matching error (system MAE).

(b) Average number of hops a resource search message will make before successfully finding the requested resource.

The ability to re-organise the contact list based on the resources information has improved the system performance by enabling the agent to learn from the system about the available resources and direct search accordingly.

4   Conclusions and Future Work

This paper has addressed resource sharing in dynamic heterogeneous Multi Agent Systems as a search problem. Performing random search might lead to traverse the whole network exhaustively and increase the failure ratio; accordingly, we have introduced heuristic directed search to overcome this problem. Our implementation of search algorithms differs from traditional algorithms by using a semantically guided technique for resource search as well as a dynamically re-organisable network of agents. The experimental results have shown that using directed search techniques is better than random search in terms of number of hops to find the match. Furthermore, network re-organisation has improved the system performance by directing the search based on resources information, especially when high accuracy is required.

We are currently working on improving the system performance by adapting the system in a way that enables it to build a virtual organisations based on the resources descriptions by employing semantic matching techniques. Furthermore, each virtual organisation will have to elect an organisation head and forward resource search messages to heads instead of individual agents which decreases the search time by traversing less agents across the network.

References

1. Al-Asfoor, M., Neville, B., Fasli, M.: A Study of the Heuristic Resource Search Algorithms. Technical Report CES-518, University of Essex, Department of Computer Science and Electronic Engineering (2012)
2. Bordini, R.H., Wooldridge, M., Hübner, J.F.: Programming Multi-Agent Systems in AgentSpeak using Jason. Wiley Series in Agent Technology. John Wiley & Sons (2007)
3. Cholvi, V., Felber, P., Biersack, E.: Efficient search in unstructured peer-to-peer networks. In: Proceedings of the Sixteenth Annual ACM Symposium on Parallelism in Algorithms and Architectures, SPAA 2004, pp. 271–272. ACM, New York (2004)
4. Klyne, G., Carroll, J.J.: Resource description framework (rdf): Concepts and abstract syntax (February 2004)
5. Ljubešić, N., Boras, D., Bakarić, N., Njavro, J.: Comparing measures of semantic similarity. In: Lužar-Stiffler, V., Dobrić, V.H., Bekić, Z. (eds.) Proceedings of the 30th International Conference on Information Technology Interfaces, pp. 675–682. SRCE University Computing Centre, Zagreb2 (2008)
6. Luke, S., Cioffi-Revilla, C., Panait, L., Sullivan, K., Balan, G.: Mason: A multiagent simulation environment. Simulation 81(7), 517–527 (2005)
7. Minar, N., Burkhart, R., Langton, C., Askenazi, M.: The swarm simulation system, a toolkit for building multi-agent simulations (1996)
8. Neville, B., Pitt, J.: PRESAGE: A Programming Environment for the Simulation of Agent Societies. In: Hindriks, K.V., Pokahr, A., Sardina, S. (eds.) ProMAS 2008. LNCS, vol. 5442, pp. 88–103. Springer, Heidelberg (2009)
9. Rao, A.S.: Agentspeak(l): Bdi agents speak out in a logical computable language. In: Proceedings of the 7th European Workshop on Modelling Autonomous Agents in a Multi-Agent World: Agents Breaking Away: Agents Breaking Away, pp. 42–55. Springer-Verlag New York, Inc., Secaucus (1996)
10. Kumar Saha, G.: Web ontology language (owl) and semantic web. Ubiquity, 1:1–1:1 (September 2007)
11. Justin Samuel, S., Sasipraba, T.: Trends and issues in integrating enterprises and other associated systems using web services. International Journal of Computer Applications 1(12), 17–20 (2010); Published By Foundation of Computer Science
12. Tisue, S., Wilensky, U.: Netlogo: A simple environment for modeling complexity. In: International Conference on Complex Systems, pp. 16–21 (2004)
13. Lin, T., Lin, P., Wang, H., Chen, C.: Dynamic search algorithm in unstructured peer-to-peer networks. IEEE Transactions on Parallel and Distributed Systems 20, 654–666 (2009)

A Quantitative Measure, Mechanism and Attractor for Self-Organization in Networked Complex Systems

Georgi Yordanov Georgiev

Department of Natural Sciences – Physics and Astronomy, Assumption College,
500 Salisbury St, Worcester MA, 01609, United States of America
ggeorgie@assumption.edu, georgi@alumni.tufts.edu

Abstract. Quantity of organization in complex networks here is measured as the inverse of the average sum of physical actions of all elements per unit motion multiplied by the Planck's constant. The meaning of quantity of organization is the number of quanta of action per one unit motion of an element. This definition can be applied to the organization of any complex system. Systems self-organize to decrease the average action per element per unit motion. This lowest action state is the attractor for the continuous self-organization and evolution of a dynamical complex system. Constraints increase this average action and constraint minimization by the elements is a basic mechanism for action minimization. Increase of quantity of elements in a network, leads to faster constraint minimization through grouping, decrease of average action per element and motion and therefore accelerated rate of self-organization. Progressive development, as self-organization, is a process of minimization of action.

Keywords: network, self-organization, complex system, organization, quantitative measure, principle of least action, principle of stationary action, attractor, progressive development, acceleration.

1 Introduction

1.1 Motivation

To define quantitatively self-organization in complex networked systems a measure for organization is necessary [1]. Two systems should be numerically distinguishable by their degree of organization and their rate of self-organization. What one quantity can measure the degree of self-organization in all complex systems? To answer this question we turn to established science principles and ask: What single principle can explain the largest number of science phenomena? It turns out that this is the principle of least (stationary) action. There is no more broad and fundamental principle in science than this, as it can be seen in the next section.

1.2 The Principle of Least Action and Its Variations

Pierre de Maupertuis stated Law of the Least Action as a "universal principle from which all other principles naturally flow" [2]. Later Euler, Lagrange, Hamilton,

F.A. Kuipers and P.E. Heegaard (Eds.): IWSOS 2012, LNCS 7166, pp. 90–95, 2012.

Fermat, Einstein, and many others refined it and applied it to develop all areas of physics [3]. It was later generalized as a path integral formalism for quantum mechanics by Feynman [4]. Jacobi's form of the principle refers to the path of the system point in a curvilinear space characterized by the metric tensor [3]. The Hertz's principle of least curvature says that a particle tends to travel along the path with least curvature, if there are not external forces acting on it [3]. The Gauss Principle of least constraint where the motion of a system of interconnected material points is such as to minimize the constraint on the system is an alternative formulation of classical mechanics, using a differential variational principle [5]. Action is more general than energy and any law derived from the principle of least action is guaranteed to be self consistent [6]. All of the laws of motion and conservation in all branches of physics are derived from the principle of least action [6,7].

1.3 Applications to Networks and Complex Systems

Scientists have applied the principle of least action to networks and complex systems. For example, it has been applied to network theory [8,9,10] and path integral approaches to stochastic processes and networks [11]. Samples of some other applications are by Annila and Salthe for natural selection [12] and Devezas for technological change [13]. Some of the other important measures and methods used in complex systems research are presented by Chaisson [14], Bar-Yam [15], Smart [16], Vidal [17] and Gershenson and Heylighen [18]. This list is not exhaustive. Some of these established measures use information, entropy or energy to describe complexity, while a fundamental quantity of physical action is used in this work to describe degree of organization through efficiency.

2 Principle of Least Action for a System of Elements – An Attractor

In a previous paper [1] we defined the natural state of an organized system as the one in which the variation of the sum of actions of all of the elements is zero. Here we define the principle of least action for n elements crossing m nodes as:

$$\delta \sum_{i=0}^{n} \sum_{j=0}^{m} I_{ij} = \delta \sum_{i=1}^{n} \sum_{j=1}^{m} \int_{t_1}^{t_2} L_{ij} dt = 0 . \tag{1}$$

Where δ is infinitesimally small variation in the action integral I_{ij} of the jth crossings between the nodes (unit motion) of the ith element and L_{ij} is the Lagrangian for that motion. n represents the number of elements in a system, m the number of motions and t_1 and t_2 are the initial and final times of each motion. $\sum_{i=0}^{n} \sum_{j=0}^{m} I_{ij}$ is the sum of all actions of all elements n for their motions m between nodes of a complex network. For example, a unit motion for electrons on a computer chip is the one necessary for one computation. For a computer network, such as internet, it is the transmission of one bit of information. In a chemical system it is the one for one chemical reaction. The state of zero variation of the total action for all motions is the one to which any system is naturally driven. Open systems never achieve this least action state because of the constant changes that occur in them, but are always tending toward it. In some respect one can consider this attractor state to be one of dynamical action equilibrium. Using the quantity of action one can measure how far the system is from this equilibrium and can distinguish between the organizations of two systems, both of which are equally close to equilibrium.

3   Physical Action as a Quantitative Measure for Organization

In [1] we defined organization of a system as inversely proportional to the average sum of all actions. Here we expand this notion by defining organization, α, as inversely proportional to the average action per one element and one motion.

$$\alpha = \frac{hnm}{\sum_{i=0}^{n} \sum_{j=0}^{m} I_{ij}} \quad .$$ (2)

h is the Planck's constant. The meaning of organization is that it is inversely proportional to the number of quanta of action per one motion of one element in a system. This definition is for a system of identical elements or where elements can be approximated as identical. It is the efficiency of physical action. The time derivative of α is the rate of progressive development of a complex system.

4   Applications

4.1   One Element and One Constraint

Consider the simplest possible part of a network: one edge, two nodes and one element moving from node 1 to node 2. Let's consider case (I) when there is no constraint for the motion of the element. It crosses the path between nodes 1 and 2 along the shortest line – a geodesic. Now consider case (II) when there is one constraint placed between nodes 1 and 2 and the shortest path of the element in this case is not a geodesic. If the path is twice as long in the second case, if the kinetic energy of the element is the same as in case (I) and no potentials are present, then the time taken to cross between nodes 1 and 2 is twice as long. Therefore the action in case (II) is twice than the action in case (I). When we substitute these numbers in the expression for organization α (eq. 2), where n=1, one element, and m=1, one crossing between two nodes, then the denominator which is just the action of the element for that motion will be twice as large in the second case and therefore the result for the amount of organization is a half as compared to the first case.

4.2   Many Elements and Constraints

Now consider an arbitrary networks consisting of three, ten, thousands, millions and billions of nodes and edges, populated by as many elements and constraints, where the paths of the elements cross each other. The optimum of all of the constraints', nodes', edges' and elements' positions and the motions of the elements is the minimum possible action state of the entire system, providing a numerical measure for its organization. Notice that action is not at an absolute possible minimum in this case, but at a higher, optimal value. Action would be at its absolute minimum only in a system without any constraints on the motion of its elements, which is not the case in complex systems and networks. Nevertheless, action is at a minimum compared to what it will be for all other arrangements of nodes, elements and constraints in the system that are less organized. When we consider an open dynamical system, where the number and positions of nodes, edges, elements and constraints constantly changes, then this minimum action state is constantly recalculated by the system. It is an attractor state which drives the system to higher level of organization and this process can continue indefinitely, as long as the system exists. Achieving maximum organization is a dynamical process in open complex systems of constantly recalculating positions of nodes, edges, elements and constraints for a least action state and preserving those positions in a physical memory of the organization of the system.

5   Exploring the Limits for Organization

5.1   An Upper Limit

The smallest possible discrete amount of action is one quantum of it, equal to the value of the Planck's constant. With self-organization the distances between the nodes shrink, the elements become smaller and the constraints for their motion decrease, for the purpose of decreasing of action (as in computer chips). The limit for this process of decrease of action is one quantum of it. If each motion uses the minimum of one quantum of action, then the value of the organization, α, is exactly one.

Can this value for organization be exceeded by a parallel processes, like quantum computing, where possibly with one, or a few quanta of action a vast number of computations can occur? Technically the crossing is still between two nodes, but it happens simultaneously along infinite number of different paths. It is like an infinite number of elements crossing between two nodes, each performing different computations. Alternatively, with decrease of the amount of action per crossing, it might be possible for the elements to cross several nodes (do several motions) with one quantum of action. In both of these cases the upper limit for organization, α, becomes very large and possibly infinity.

5.2   A Lower Limit

For a completely disorganized system, where the entropy is at a maximum, all points in the system are equally probable for an element to visit. In order to reach its final destination, an element of the system must visit all points in it (by definition for maximum entropy), thus maximizing its action for one crossing from any node 1 to any node 2. In this case, the action is extremely large and the organization, α, of this system is very close to zero.

Another way to describe the lower limit for organization of a system is when the constraint for the motion between nodes 1 and 2 is infinitely large and the path taken by the element to cross between the nodes is infinitely long. This also maximizes action and describes a completely disorganized system. The value for organization, α, in this case again approaches a limit of zero.

6   Mechanism of Self-Organization

When elements interact with constraints they apply force to minimize them, lowering their action for the next cycle. With the increase of quantity in a system, several elements can group on the same constraint to minimize it for less time. Decreased average action makes a system more stable, by lowering the energy needed for each motion. High average action, in disorganized system destabilizes it and above some limit it falls apart. Therefore a system with low enough average action can increase its quantity within limits of stability. Quantity and level of organization are proportional. If the quantity becomes constant, then the organization will reach a least action state and stop increasing. For continued self-organization an increase of the quantity is necessary. Quantity and level of organization of a system are in an accelerating positive feedback loop, ensuring unlimited increase of the level of organization in a system, unless it is destroyed by external influence, like limited resources, huge influx of energy, force impact, change in the conditions, etc.

7   Conclusions

The principle of least action for a networked complex system (eq. 1) drives self-organization in complex systems and the average action is the measure of degree to which they approach this least action state. Actions that are less than their alternatives are self-selected. Progressive development, as self-organization, is a process of minimization of action. In open systems there is a constant change of the number of elements, constraints and energy of the system and the least action state is different in each moment. The process of self-organization of energy, particles, atoms, molecules, organisms, to the today's society is a process of achieving a lower action state, with the least action as a final state. The laws of achieving this least action state are the laws of self-organization. The least possible action state is the limit for organization when time is infinite and all elements in the universe are included.

The state of nodes, edges, constraints and elements that determines the action for one motion in a system is its organization. With its measure α (eq. 2) we can compare any two systems of any size and the same system at two stages of its development. It distinguishes between systems with two different levels of organization and rates of self-organization and is normalized for their size. The measure can be applied to all systems and researchers in all areas studying complex systems can benefit from it. With a quantitative measure we can conduct exact scientific research on self-organization of complex systems and networks, progressive development, evolution and co-evolution, complexity, etc.

Acknowledgments. The author thanks Assumption College for support and encouragement of this research, and Prof. Slavkovsky, Prof. Schandel, Prof. Sholes, Prof. Theroux and Prof. Cromarty for discussion of the manuscript. The author thanks Francis M. Lazarus, Eric Chaisson, John Smart, Clement Vidal, Arto Annila, Tessaleno Devezas, Yaneer Bar-Yam and Atanu Chatterjee for invaluable discussions.

References

1. Georgiev, G., Georgiev, I.: The least action and the metric of an organized system. Open Syst. Inf. Dyn. 9(4), 371 (2002)
2. de Maupertuis, P.: Essai de cosmologie (1750)
3. Goldstein, H.: Classical Mechanics. Addison Wesley (1980)
4. Feynman, R.: The Principle of Least Action in Quantum Mechanics. Ph.D. thesis (1942)
5. Gauss, J.: Über ein neues allgemeines Grundgesetz der Mechanik (1831)
6. Taylor, J.: Hidden unity in nature's laws. Cambridge University Press (2001)
7. Kaku, M.: Quantum Field Theory. Oxford University Press (1993)
8. Hoskins, D.A.: A least action approach to collective behavior. In: Parker, L.E. (ed.) Proc. SPIE, Microrob. and Micromech. Syst., vol. 2593, pp. 108–120 (1995)
9. Piller, O., Bremond, B., Poulton, M.: Least Action Principles Appropriate to Pressure Driven Models of Pipe Networks. In: ASCE Conf. Proc. 113 (2003)
10. Willard, L., Miranker, A.: Neural network wave formalism. Adv. in Appl. Math. 37(1), 19–30 (2006)
11. Wang, J., Zhang, K., Wang, E.: Kinetic paths, time scale, and underlying landscapes: A path integral framework to study global natures of nonequilibrium systems and networks. J. Chem. Phys. 133, 125103 (2010)
12. Annila, A., Salthe, S.: Physical foundations of evolutionary theory. J. Non-Equilib. Thermodyn., 301–321 (2010)
13. Devezas, T.C.: Evolutionary theory of technological change: State-of-the-art and new approaches. Tech. Forec. & Soc. Change 72, 1137–1152 (2005)
14. Chaisson, E.J.: The cosmic Evolution. Harvard (2001)
15. Bar-Yam, Y.: Dynamics of Complex Systems. Addison Wesley (1997)
16. Smart, J.M.: Answering the Fermi Paradox. J. of Evol. and Tech. (June 2002)
17. Vidal, C.: Computational and Biological Analogies for Understanding Fine-Tuned Parameters in Physics. Found. of Sci. 15(4), 375–393 (2010)
18. Gershenson, C., Heylighen, F.: When Can We Call a System Self-Organizing? In: Banzhaf, W., Ziegler, J., Christaller, T., Dittrich, P., Kim, J.T. (eds.) ECAL 2003. LNCS (LNAI), vol. 2801, pp. 606–614. Springer, Heidelberg (2003)

MetroNet: A Metropolitan Simulation Model Based on Commuting Processes

Efrat Blumenfeld-Lieberthal[1] and Juval Portugali[2]

[1] The Azrieli School of Architecture, Tel Aviv University, Israel
efratbl@post.tau.ac.il
[2] The Department of Geography and Human Environment, Tel Aviv University, Israel
juval@post.tau.ac.il

Abstract. The aim of this work is to identify a set of fundamental rules that govern the interactions within urban systems at the metropolitan scale. For that, we developed an USM (Urban Simulation Model) specifically designed to study the evolution and dynamics of systems of cities. Our model is innovative in its structure: it is a superposition of cellular automata and agent based modeling approaches (that are essentially spatial analyses) and a complex network approach (that is essentially a topological analysis). This implies that in our model, the local activities and interaction of agents give rise to the global urban structure and network that in turn affects the agents' cognition, behavior, movement and action in the city and so on in circular causality. The model simulates commuting patterns of agents within a metropolis. The agents in our model represent workers who look for working places, the nodes represent urban employment centers, and the links represent commuters. Our results address three issues: the first suggests that the perception of urban boundaries plays a significant role in the metropolitan evolution in terms of network topology. This means that the existence of business centers, located in proximity to each other (but belonging to different municipalities) may lead to the emergence of new centers at the metropolis scale. The second issue concerns urban segregation; our results suggest that the location preferences of the agents regarding proximity to similar/different agents have a major affect not only on the urban morphology but also on the topology of the urban network. The third and last issue concerns the size distributions of agents in our model; these distributions correspond to all types of homogenous distributions observed in real system of cities.

Keywords: urban complexity, urban networks, agent based models, urban simulation models.

1   Introduction

In 1965, Christopher Alexander in his paper "A city is not a tree" described the differences between the way modernist planners grasp the relationships between urban entities and the way they exist in reality. A careful reading of this work suggests that Alexander was one of the first scholars that proposed a new understanding of cities as

F.A. Kuipers and P.E. Heegaard (Eds.): IWSOS 2012, LNCS 7166, pp. 96–103, 2012.

complex systems that can be described as complex networks. These networks contain many entities that interact in many levels and scales. According to Alexander, they must be treated as a system rather than separate objects that act within the urban context.

When looking into cities, it always comes down to urban agents that interact with each other. These agents can be individual human beings, families, households, organizations, municipalities, firms, and so on. They interact in a variety of domains such as economic, infrastructures, transportation facilities, commuting, internal immigration and more.

Since the works of Watts and Strogatz [1] and of Barabasi and Albert [2], the science of complex networks has been developing rapidly. It has been applied to various disciplines in order to study the topology of large networks and to understand their development and robustness [3-11]. Most of these works, however, consider the topological characteristics of urban networks, while the physical aspect is mostly neglected.

Cities are typical examples of complex, self-organizing systems [10-12]. They have originally emerged and are still developing out of the interactions between many agents that are located and move in space and time [13, 14]. These agents are motivated by a variety of forces ranging from cognitive capabilities and needs to economic considerations, political ambitions, etc., with no central force that affects their behavior. These interactions entail a huge number of links that create complex networks which form the city.

Most if not all complexity theories and models have been applied to the study of cities with the implication that we now have a whole family of urban simulation models that model cities (See review in [11, 15]). In the last decade or so cellular automata and agent-base models (CA/AB) have become the main medium to simulate cities as complex systems [15-17] while in the last few years we see studies that model cities as complex networks that are often typified by power law distributions [12, 18-19].

In this work, we focus on urban networks and study their topological as well as spatial characteristics. We aim to identify a set of fundamental rules that govern the interactions within urban systems at the metropolitan scale. For that, we developed a USM (Urban Simulation Model) specifically designed to study the evolution and dynamics of systems of cities. Our model is innovative in its structure as it is a superposition of cellular automata and agent based modeling approaches (that are essentially spatial analyses) and a complex network approach (that is essentially a topological analysis). This implies that in our model, the local activities and interaction of agents give rise to the global urban structure and network that in turn affects the agents' cognition, behavior, movement and action in the city and so on in circular causality.

In the next section we present the detailed description of the above model. Then, we present some preliminary results. As the presented work is an ongoing one, in the last section, we elaborate on directions for future work.

2   A Detailed Description of the Model

The essence of our model can be described by the following scenario: we start with a metropolitan environment divided into sub-areas, some of which represent cities with their municipal boundaries and others represent green areas. Each city is growing logistically (e.g. demographically due to natural increase and/or migration), and economically due to the development of employment and business centers. The residents (agents) of the city seek working places either in their town of residence or in other cities (to which they need to commute). Their decision where to work in based on a gravitation spatial interaction model i.e. their choice of a preferred working place is proportional to the size of the existing employment centers and inversely proportional to the distance between the agent's home and employment center. The choices where to work give rise to an intra- and inter-city commuting system. When a certain threshold of commuters between two cities is crossed, a link between these cities is created. The weight of the link is a dynamic parameter, which represents the volume of commuters.

Our USM is composed of the following elements and process:

1. The metropolis infrastructure:
   (a) The area of the metropolis which is represented by a rectangular is defined (by the user)
   (b) The model divides the metropolis into 25 spatial equal units and characterizes 7 of them (based on a random choice) as green areas (where urban development is restricted)
   (c) The remaining 18 urban units are randomly united into 12 municipalities. These numbers are comparable to the Tel Aviv metropolis (to which we intend to relate this work in later stages).
   (d) For each municipality, a center is randomly selected and being occupied by one of the agents. This random choice of the urban center can be explained as a historical accident [12].
2. Characterizing up to 3 types of agents. Each type of agents is represented by:
   (a) Growth rate - the addition of new agents in each iteration
   (b) Sensitivity to size of employment opportunities (α – see below)
   (c) Sensitivity to distance between the agent's place of residence and work (β – see below)
3. Location method:
   (a) At the first stage: for each municipality an agent is randomly selected and located. Each new agent can be located either in its original city or elsewhere based on the availability of vacant space and a probability function. If the agent is not located in its original town its new location is determined by a **gravitation function**. The gravitation function can be described by: $G = \frac{(Mm)^{\alpha}}{r^{\beta}}$, where G represents the gravitation force, M and m represent the size (total number of agents) in the agent's original and destination towns (correspondingly), r represents the physical distance between the centers of both towns, and α and β are parameters that characterize the sensitivity of the agent to size and distance correspondingly.
   (b) The segregation function can be turned On/Off. This determines whether agents want to locate close to agent of their type.
4. Network definition:
   (a) Nodes are defined at the location of the first agent in each city. The historical center is considered as the city's core

(b) A minimum volume of commuters (agents) is needed to define a link. This number is predefined as a percentage of the population (and is changeable)
(c) The weight of the links is presented (gradient color change)
(d) When a predefined percentage of the cells, located in proximity to the municipal boundaries are occupied by agents, an agglomeration emerges. This means that the municipal boundaries are ignored in the gravitation function and the new center of the agglomeration is moved to its geographical center. Note that this condition is ignored when the predefined percentage of cells, located near the municipal boundary is 100.

3   Preliminary Results

Based on the above USM, we can introduce some interesting preliminary results. In figures 1-2 (bottom), the green rectangles represent green areas where urban development is restricted, while the black lines represent municipal boundaries. The different types of agents are represented by black, red, and grey pixels. The first issue we address concerns agglomerations in face of municipal boundaries. Figure 1 presents the morphology and the network topology of the metropolis in two different runs of the model. In both runs there is only one type of agents. In the first run of the model (figure 1a), the agents consider the municipal boundaries in their choice of working places, while in the second run (figure 1b) the agents consider agglomerations instead of municipalities in their choice of working places. It can be seen that the perception of urban boundaries plays a significant role in the metropolitan evolution in terms of network topology. While in figure 1a (top) the topology of the network is a star network, in figure 1b (top) the star network disappears and a new topology emerges. This new topology suggests that the existence of business centers, belonging to different municipalities but located in proximity to each other, might lead to the emergence of new centers at the metropolitan scale. In other words, employment centers at the metropolitan scale can span beyond the boundaries of a single municipality.

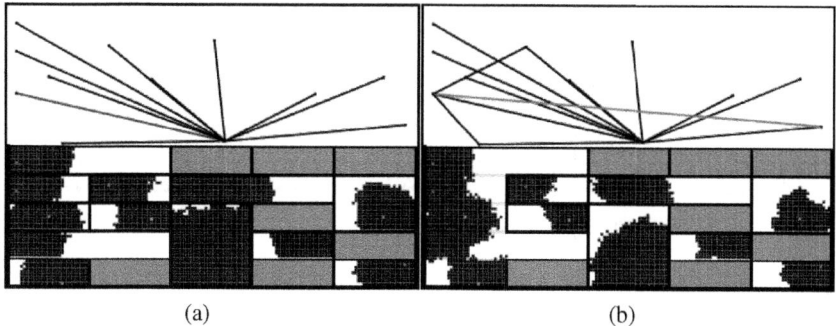

(a)   (b)

Fig. 1. The model at the metropolitan scale with one type of agents: a) municipal boundaries are considered and b) agglomerations of built-up areas replace municipalities

The second issue concerns urban residential segregation. Figure 2 presents the morphology and network topology of the metropolitan in two different runs of the model in which there are 3 different types of agents. In the first run the agents are indifferent to the kind of their neighbors, thus, they might be located in adjacency to other kinds of agents (figure 2a, bottom). In the second run the agents prefer to locate in proximity to agents of their own kind (figure 2b bottom). To control this behavior, we added a segregation function to the model. When in use, this function affects the gravitation function such that M and m are calculated based only on the number of agents of the same kind as the agent who looks for a location.

It can be seen that the location preferences of the agents regarding proximity to similar/different agents have a major affect both on the urban morphology and network topology. When the agents prefer to locate in proximity to their own kind – urban segregation emerges. In terms of network topology, a significant difference between both runs is observed. When plotting the rank-size distribution of the links' weight (Y=log(weight of link) and X=log(rank of link)) the distribution of the first run (agents have no preferences of neighbors) corresponds to an exponential function (figure 3a), while the distribution of the second run (agents prefer to work in proximity to their own kind) obeys a power law (figure 3b).

In addition, in the first run the highest level of commuters reaches approximately 11% of the total population in the metropolis. On the other hand, in the second run the highest level of commuters reaches only 6.7% of the total population. This can be explained by the fact that in the first run, the large employment centers attracted all three types of agents. However, in the second run, employment centers attracted only agents of the same type as their majority population.

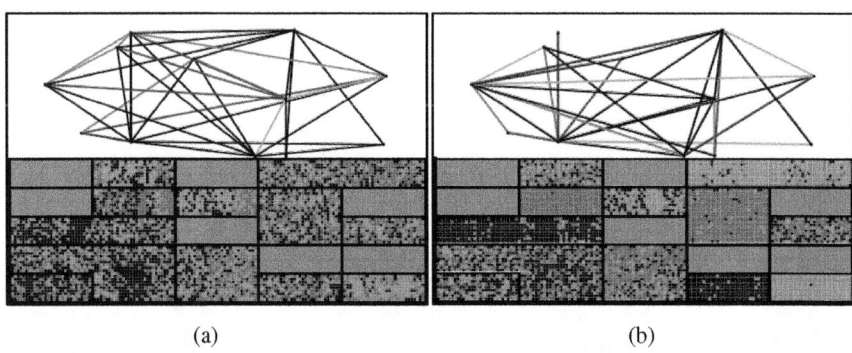

(a)   (b)

Fig. 2. The model at the metropolitan scale with 3 types of agents: a) agents are tolerant to location in proximity to other types of agents b) agents prefer to locate in proximity to their own types

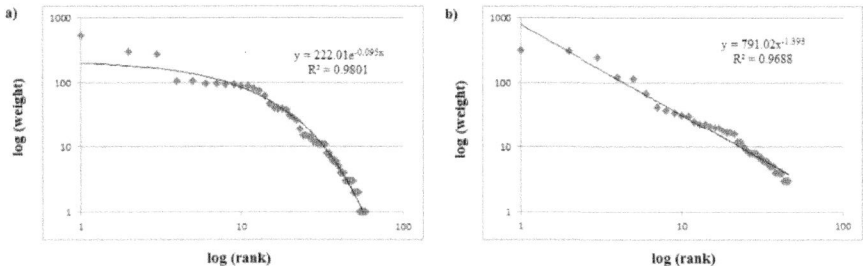

Fig. 3. Rank size distribution of links' weight: a) when agents are tolerant to location in proximity to other types of agents and b) when agents prefer to locate in proximity to their own types

Lastly, we address the rank-size distribution of the agents in our model. There is an extensive work on scaling relationship in cities [20-25]. One of the classic works is that of Haggett [26] who suggested that city size distributions of urban systems can be divided into three classes: the first includes system of cities with primate city (the largest city is considerably larger than the next largest one), the second class includes power low distributions and the third includes systems of cities in which the large cities are rather homogenously distributed. An empirical study on the size distribution of cities [25] showed that homogenous systems of cities can indeed be classified into these classes. When analyzing the resulted size distributions of the different agents in our model, we get all the observed population size distributions (see examples in figure 4). This, of course corresponds to the different values of parameters we used in different runs. This is also valid to the network topology. We found that the rank size distributions of the links' weight also correspond to these three classes. These results suggest that the set of laws to which the agents in our model obey, correspond to the behavior of people in real systems of cities. By following these laws and the values of the parameters that represent them we can try and understand which forces are the most dominant in the creation of scaling relationship within urban systems (as opposed to other parameters that lead to other relations).

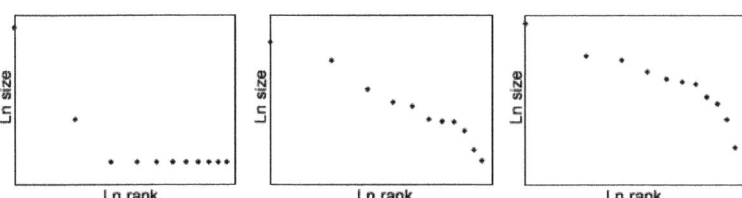

Fig. 4. The three classes of city size distributions, recovered by different runs of our model (with different parameters at each run)

4   Discussion and Conclusion

In this paper we've presented an urban simulation model designed to study the evolution and dynamics of metropolitan systems of cities. Our model is built as a superposition of cellular automata and agent based modeling approaches (that are essentially spatial analyses) and a complex network approach (that is essentially a topological analysis). We study both the topology and the spatial characteristics of urban commuting networks at the metropolis scale.

In a future work, we intend to elaborate our model in order to study additional urban phenomena; the first addresses the morphology of the municipal boundaries. To explore this issue we will change the definition of municipal boundaries and study the effect of increased irregularity of the boundaries' morphology on the resulted networks. The second phenomenon addresses preservation of rural environment under the economic pressures of real-estates developers. In reality rural areas are often transformed into developed urban environments (e.g. edge cities). In our next version of the model, we will enable (under controlled conditions) urban development in the rural (green) areas and study its effect on the urban networks in terms of the networks' topology and spatial characteristics.

References

1. Watts, D.J., Strogatz, S.H.: Collective dynamics of 'small-world' networks. Nature 393(6684), 409–410 (1998)
2. Barabási, A.L., Albert, R.: Emergence of Scaling in Random Networks. Science 286(5439), 509–512 (1999)
3. Albert, R., Jeong, H., Barabási, A.L.: Error and attack tolerance of complex networks. Nature 406, 378–382 (2000)
4. Gallos, L.K., Cohen, R., Argyrakis, P., Bunde, A., Havlin, S.: Stability and topology of scale-free networks under attack and defense strategies. Phys. Rev. Lett. 94, 188701 (2005)
5. Paul, G., Sreenivasan, S., Havlin, S., Stanley, H.E.: Optimization of network robustness to random breakdowns. Physica A 370, 854–862 (2006)
6. Jeong, H., Tombora, B., Albert, R., Oltvai, Z.N., Barabási, A.L.: The large-scale organization of metabolic networks. Nature 407, 651–654 (2000)
7. Newman, M., Watts, D.J., Strogatz, S.H.: Random graph models of social networks. PNAS 99, 2566–2572 (2002)
8. Jeong, H., Neda, Z., Barabási, A.L.: Measuring preferential attachment in evolving networks. Europhys. Lett. 61, 567–572 (2003)
9. Blumenfeld-Lieberthal, E.: The topology of transportation networks: a comparison between different economies. Networks and Spatial Economics 9, 427–458 (2009)
10. Patuelli, R., Reggiani, A., Nijkamp, P., Bade, F.J.: The evolution of the commuting network in Germany: spatial and connectivity patterns. Journal of Transport and Land Use 2(3) (2010)
11. Batty, M.: Cities and complexity. The MIT Press, Cambridge (2005)
12. Porutgali, J.: Complexity, Cognition and the City. Springer, Heidelberg (2011)
13. Portugali, J.: Self-Organization and the City. Springer, Berlin (2000)

14. Alexander, C.: A city is not a tree. Architectural Forum 122(1), 58-61 and 122(2), 58-62 (1965)
15. Batty, M., Longley, P.: Fractal Cities: A Geometry of Form and Function. Academic Press, London (1994)
16. Benenson, I., Torrens, P.: Geosimulation: Automata-based modeling of urban phenomena. John Wiley and Sons Ltd., West Sussex (2004)
17. White, R., Engelen, G.: Cellular automata and fractal urban form: a cellular modeling approach to the evolution of urban land-use patterns. Environment and Planning A 25, 1175–1199 (1993)
18. Andersson, C., Frenken, K., Hellervik, A.: A complex network approach to urban growth. Environment and Planning A 38, 1941–1964 (2006)
19. Porta, S., Crucittib, P., Latora, V.: The network analysis of urban streets: A dual approach. Physica A 369, 853–866 (2006)
20. Jiang, B., Claramunt, C.: Topological analysis of urban street networks. Environment and Planning B 31(1), 151–162 (2004)
21. Zipf, G.K.: National unity and disunity. The Principia Press, Bloomington Indiana (1941)
22. Krugman, P.: The Self-Organizing Economy. Blackwell, Cambridge (1996)
23. Batty, M.: Rank clocks. Nature 444, 592–596 (2006)
24. Geoffery, F., Linsker, R.: Synchronous neural activity in scale-free network models versus random network models. PNAS 102(28), 9948–9953 (2005)
25. Benguigui, L., Blumenfeld-Lieberthal, E.: A new classification of city size distributions. Computers, Environment and Urban Systems 31(6), 648–666 (2006)
26. Haggett, P.: Models in geography: the Madingley lectures for 1965. Edited with Chorley, R.J., Methuen & Co. Ltd., London (1967)

Robustness of Self-Organizing Consensus Algorithms: Initial Results from a Simulation-Based Study

Alexander Gogolev[1] and Christian Bettstetter[1,2]

[1] Institute of Networked and Embedded Systems, University of Klagenfurt, Austria
[2] Lakeside Labs GmbH, Klagenfurt, Austria

Abstract. This short paper studies distributed consensus algorithms with focus on their robustness against communication errors. We report simulation results to verify and assess existing algorithms. Gacs-Kurdyumov-Levin and simple majority rule are evaluated in terms of convergence rate and speed as a function of noise and network topology.

Keywords: Networked systems, self-organization, consensus, robustness.

1 Introduction

The overall objective of our research is to analyze *robustness* characteristics of *features of self-organization* in networks. We recognize any feature of a network as feature of self-organization if it is capable of performing actions aimed at maintaining the network function in a completely distributed manner. Each node has only local view, and simple rules are applied. Systems presenting such features are recognized as "having an ability of self-organization." We use the term robustness as ability of the system (or an algorithm as part of the system) to maintain a certain level of network function despite changes or noise in the environment and/or changes in system structure, including partial failure.

This short paper is our first step in this research direction. It studies self-organizing *consensus* algorithms in networks with focus on robustness toward communication errors between nodes. Consensus algorithms can be applied to a broad spectrum of tasks, ranging from distributed version management over analysis of social networks to forecasting of crowd behavior.

Our work is inspired by an article of Moreira *et al.* [1]. That paper studies the impact of noise leading to communication errors on distributed consensus. Gacs-Kurdyumov-Levin (GKL) [2] and simple majority rule (SMR) [3,4] are tested for convergence under different noise levels in a small-world network [5]. It was shown that noise can actually improve performance of simple algorithms in certain setups [1]. The work at hand slightly extends that work using almost the same modeling assumptions but investigating a larger set of parameters.

F.A. Kuipers and P.E. Heegaard (Eds.): IWSOS 2012, LNCS 7166, pp. 104–108, 2012.

2   Modeling Assumptions

2.1   Setup and Notation

The system is modeled as a cellular automata graph [6]. Given is a one-dimensional grid of n nodes where each node is connected to its k closest neighbors. At a time $t = 0$, nodes are initially assigned with a random binary state σ whose value is either $+1$ (white color) or -1 (black color). At every time step $t > 0$, all nodes update their state following a given consensus rule based on their own state and their neighbors' states. The network is expected to converge within $2n$ time steps to a single value, which corresponds to the initial majority of value distribution [1]. The state of node i at time t is called $\sigma_i[t]$; the state that node i receives for the state of node j is called $\tilde{\sigma}_j^i[t]$.

2.2   Consensus Algorithms

Let us describe the GKL and SMR rules. According to GKL, a node i computes its new state $\sigma_i[t+1]$ using its current state $\sigma_i[t]$ and the current state of the first and third nearest neighbors to the left or right from the node. The current value defines the side from which the neighbors are chosen. We have [1]

$$\sigma_i[t+1] = \begin{cases} G\left(\tilde{\sigma}_i^i[t] + \tilde{\sigma}_{i-1}^i[t] + \tilde{\sigma}_{i-3}^i[t]\right) & \text{if } \tilde{\sigma}_i^i[t] = -1 \\ G\left(\tilde{\sigma}_i^i[t] + \tilde{\sigma}_{i+1}^i[t] + \tilde{\sigma}_{i+3}^i[t]\right) & \text{if } \tilde{\sigma}_i^i[t] = +1 \end{cases}, \qquad (1)$$

where the nodes $i-1$ and $i-3$ are the first and third neighbors to the left (nodes $i+1$ and $i+3$ are to the right). The update function G is [1]

$$G(x) \equiv \begin{cases} -1 & \text{for } x < 0 \\ +1 & \text{for } x > 0 \end{cases}. \qquad (2)$$

Using SMR, a node calculates its new value on basis of its current state and the state of its closest k neighbors from both sides. i.e.,

$$\sigma_i[t+1] = G\left(\sum_{j=i-k}^{i+k} \tilde{\sigma}_j^i[t] \right). \qquad (3)$$

2.3   Noise

Noise is quantified by a noise parameter $\eta \in [0,1]$. Noise-free environments are modeled using $\eta = 0$; random dynamics is $\eta = 1$ [1]. Communication errors caused by noise are then modeled as follows [1]:

$$\tilde{\sigma}_j^i[t] = \begin{cases} \sigma_j[t] & \text{with probability } (1 - \eta/2) \\ -\sigma_j[t] & \text{with probability } \eta/2 \end{cases}. \qquad (4)$$

3   Performance Study

3.1   Contributions and Methodology

We extend some parts of the results reported in [1] by using a larger set of system parameters. As network topologies, we apply the commonly used Watts-Strogatz and Newman-Watts-Strogatz models [7] on a one-dimensional grid and vary the number of neighbors from 2 to 5. The noise parameter $\eta/2$ is varied from 0 to 1. Moreover, we consider $k = 2\ldots5$ neighbors in SMR. Every set of conditions is run 1 000 times with both synchronous and asynchronous update functions. The simulation engine is built with the programming language Python and uses the NetworkX software package (`networkx.lanl.gov`).

As performance metrics we measure the convergence rate (fraction of converged networks out of 1 000 simulations) and convergence speed (number of steps until network converges to a single state normalized by the overall number of steps within one simulation).

There appear "noise artifacts" at certain noise levels. These artifacts are ignored in [1], i.e., a network is considered as converged even if not all nodes have the same value. We only consider the consensus to be converged if really all nodes have the same state. Therefore our simulations assume somewhat more difficult conditions.

3.2   Simulation Results

For illustration, Figs. 1 and 2 show the evolution of the nodes' states over time for given starting conditions. These plots basically resemble previous results [1].

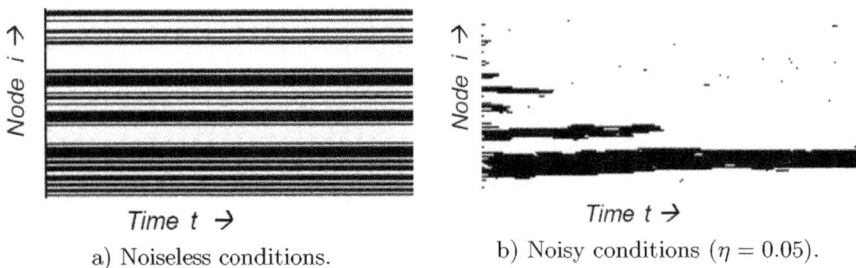

a) Noiseless conditions.   b) Noisy conditions ($\eta = 0.05$).

Fig. 1. SMR rule in a network of $n = 100$ nodes with Watts-Strogatz connectivity and zero rewiring probability. Both networks do not convergence in the given time. a) Clustering with stable domain boundaries. b) Clustering with unstable domain boundaries.

Figs. 3 and 4 show the convergence rate and speed with different parameters and network models. The curves can be interpreted as follows:

– Using GKL, low and medium noise levels lead to a convergence rate of about 50%, while no and high noise tend to prevent consensus (Fig. 3a).

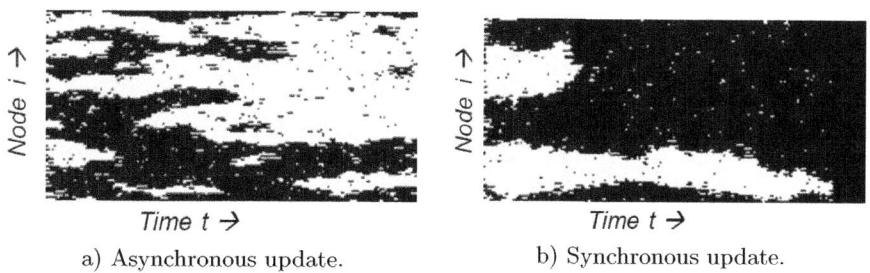

a) Asynchronous update.   b) Synchronous update.

Fig. 2. GKL rule under noisy conditions in a network of $n = 100$ nodes with Watts-Strogatz connectivity and zero rewiring probability. a) No convergence. b) Convergence.

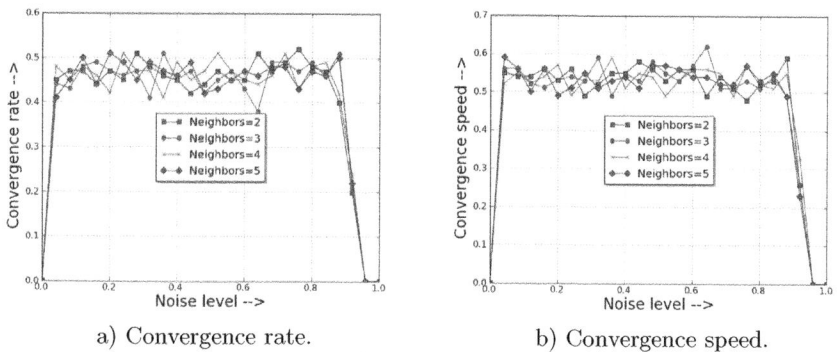

a) Convergence rate.   b) Convergence speed.

Fig. 3. GKL rule in a network of $n = 100$ nodes with Watts-Strogatz connectivity and rewiring probability 0.5 for topologies with different number of neighbors

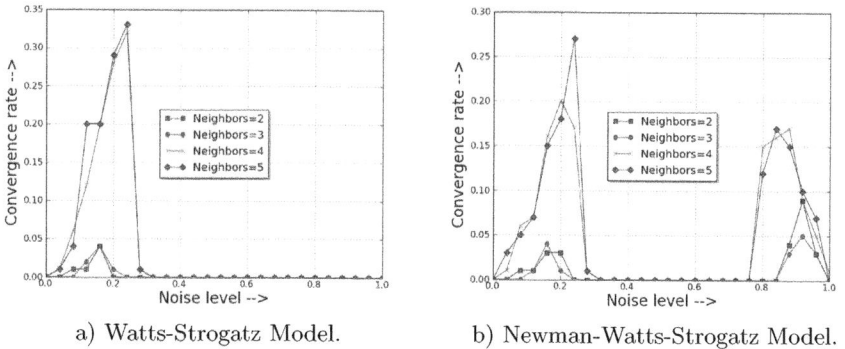

a) Watts-Strogatz Model.   b) Newman-Watts-Strogatz Model.

Fig. 4. SMR rule in a network of $n = 100$ nodes with rewiring probability 0.9

- With Newman-Watts-Strogatz networks (Fig. 4b), there is an extra peak in convergence rate for high noise levels using SMR with synchronous update.[1]
- Both GKL and SMR have weaker performance than in [1], which is due to the fact that we did not mitigate "noise artifacts" in measuring convergence.

In future work, the simulation engine will be used for robustness analysis with respect to other types of faults and in two-dimensional topologies.

Acknowledgments. This work was performed within the Erasmus Mundus Joint Doctorate in "Interactive and Cognitive Environments," which is funded by the EACEA Agency of the EC under EMJD ICE FPA n 2010-0012. The work of A. Gogolev is supported by Lakeside Labs, Klagenfurt, with funding from the ERDF, KWF, and the state of Austria under grant 20214/21530/32606.

References

1. Moreira, A.A., Mathur, A., Diermeier, D., Amaral, L.: Efficient system-wide coordination in noisy environments. Proc. National Academy of Sciences of the USA 101(33), 12085–12090 (2004)
2. Gacs, P., Kurdyumov, G.L., Levin, L.A.: One dimensional uniform arrays that wash out finite islands (in Russian). Problemy Peredachi Informatsii 14, 92–98 (1978)
3. Boyd, R., Peter, J.R.: Why does culture increase human adaptability? Ethology and Sociobiology 16, 125–143 (1995)
4. Heyes, C.M., Galef Jr., B.G. (eds.): Social Learning in Animals: The Roots of Culture. Academic Press, San Diego (1996)
5. Watts, D.J.: Small Worlds: The Dynamics of Networks Between Order and Randomness. Princeton University Press (1999)
6. Wolfram, S.: A New Kind of Science. Wolfram Media, Champaign (2002)
7. Newman, M.: The structure and function of complex networks. SIAM Review 45(2), 167–256 (2003)

---

[1] As the Newman-Watts-Strogatz model represents social networks and excessive noise might correspond to a crowd-like message delivery, these extra peak could maybe be interpreted as potential ability of achieving consensus in social group with faulty message delivery.

Initial Experiments
in Using Communication Swarms
to Improve the Performance of Swarm Systems

Stephen M. Majercik

Bowdoin College, Brunswick ME 04011, USA

Abstract. Swarm intelligence can provide robust, adaptable, scalable solutions to difficult problems. The distributed nature of swarm activity is the basis of these desirable qualities, but it also prevents swarm-based techniques from having direct access to global knowledge that could facilitate the task at hand. Our experiments indicate that a swarm system can use an auxiliary swarm, called a *communication swarm*, to create and distribute an approximation of useful global knowledge, without sacrificing robustness, adaptability, and scalability. We describe a communication swarm and validate its effectiveness on a simple problem.

1 Introduction

Swarm intelligence is a natural phenomenon in which complex behavior emerges from the collective activities of a large number of simple individuals who interact with each other and sense the environment only in their immediate area. They have limited memory and a limited repertoire of simple, reactive behaviors. Yet, the swarm as a whole is highly organized and complex behavior emerges at the global level that goes beyond the capabilities of the individuals.

We are interested in artificial *task-swarms*, i.e. swarms that have been designed by people to accomplish a particular task. As a motivating example, consider a swarm of unmanned aerial vehicles engaged in reconnaissance. An independent swarm might be assigned the task of protecting them from attacks by hostile swarms, and it would be helpful for the members of this swarm to have information about the size of their swarm relative to that of the attacking swarm in order to adapt their behavior to their current size (dis)advantage. But, the distributed nature of swarm activity, although providing the advantages of adaptability, robustness, and scalability, also prevents swarm members from having direct access to *global knowledge*, i.e. knowledge about the state of the entire swarm and/or environment (including other swarms). Since the nature of a task-swarm precludes the possibility of its members having perfect global knowledge in real time, we will be concerned with the creation and distribution of *pseudo-global knowledge* (PGK), or knowledge about the state of the system that may be imperfect because it is only an approximation of the true state of the system, or because the state being approximated is a past state.

F.A. Kuipers and P.E. Heegaard (Eds.): IWSOS 2012, LNCS 7166, pp. 109–114, 2012.

PGK needs to be supplied to a task-swarm in a way that does not sacrifice robustness and scalability, ruling out any approaches that depend on a small number of specific bots to create and deliver the information. Task-bots could attempt to construct PGK based on environmental information they collect and information they receive from other task-bots, but this may be difficult for four reasons. First, the bots' task may limit their contact with other bots and the fraction of the environment they visit, such that they are unable to aggregate enough information in a short enough time span for the information to be useful. Second, their task may interfere with the information collection process. Third, as the scale of aerial bots approaches the micro or even nano level, their capabilities might be too limited to allow them to do their task and engage in the communication that would be necessary to construct PGK. Finally, very small-scale bots might have only enough power for very short range communication, making the construction of PGK difficult. Stigmergic communication, in which individuals communicate indirectly by altering the environment, is an alternative means of communication. This type of communication has been implemented using digital pheromones in [2], but that approach relies on a network of grounded sensors. Lack of space prevents a detailed literature review.

We propose to address these issues by using a specialized, auxiliary swarm, called a *communication swarm*, whose members, called *comm-bots*, create PGK and distribute that knowledge to other swarms. A communication swarm, or *comm-swarm*, operates in a swarm-like, distributed fashion, and so preserves the adaptability, robustness, and scalability of the entire system.

In Section 2, we introduce communication swarms using a simple illustrative problem and present the results of some initial experiments that validate the idea of communication swarms. We discuss possible future work in Section 3.

2   The FLY-HOME Problem and Experimental Results

The operation of a comm-swarm will depend on the nature of the PGK it is providing; the FLY-HOME problem illustrates the operation of a particular comm-swarm. In this problem, the bots in each of two task-swarms need to determine whether their swarm has fewer members than the other swarm and, if so, to fly to a specified location. The specific action triggered is not critical; the essential idea is that the PGK allows the swarm to act appropriately. We note that this problem would be extremely difficult for task-bots making decisions based on information gathered locally, even over an extended time period, since they would find it impossible to determine, if surrounded by their fellow task-bots over a period of time, whether they were actually in the majority, or were in the minority but were part of a congregation around the home location. Comm-swarms provide a solution to this problem.

The algorithm governing task-bot motion is similar to Reynolds' boids algorithm [3]. Each bot tries to maintain a specified speed, subject to a maximum, while staying close to its neighbors (cohesion), matching their average velocity (alignment), and keeping a distance from them (separation). Neighbors are those bots that are within a circle of a specified radius. Weights specify the strength of separation (0.0–100.0), cohesion (0.0–1.0), and alignment (0.0–1.0). Time proceeds in discrete steps and, at each step, the factors just described are used to update a bot's velocity, which is then applied to calculate its new position.

We give an overview of our comm-swarm based algorithm for the FLY-HOME problem; lack of space prevents details. Both task-bots and comm-bots maintain cumulative counts of the number of bots they have encountered in each task-swarm. At each time step: 1) each comm-bot increases its cumulative counts by the number of task-bots from each swarm in its neighborhood, 2) each comm-bot's cumulative counts are discounted by a factor of 0.95 to allow it to gradually "forget" past counts and adapt to changes in swarm sizes more quickly, 3) each task-bot increases its cumulative counts by the cumulative counts of each comm-bot in its neighborhood, and 4) each task-bot decides whether to fly home based on its counts. If a task-bot's count of bots in its swarm is less than its count of bots in the other swarm, it flies home *at that time step only* by 1) changing its motion parameters such that it coheres and aligns more strongly with bots from its swarm and enforces less of a separation from them, making it possible for the task-bots flying home to congregate closely around the home area, and 2) adjusting its velocity to include a component toward the home position. If a task-bot's count of bots in its swarm is greater than or equal to its count of bots in the other swarm, it uses its original behavioral parameters to update its velocity and position at that time step, possibly interrupting its flight home.

In our experiments, Task-Swarms 1 and 2 were initialized with 500 and 1000 bots, respectively, and the comm-swarm was initialized to the comm-swarm size being tested, all bots randomly distributed in a 2500 pixel × 1350 pixel non-toroidal environment. (Although aerial swarms would operate in three dimensions, we used a 2-dimensional version of the problem to test the comm-swarm idea.) The home location was the center of the environment, and a swarm was defined to be at home if at least 80% of its bots were in a 400 pixel × 400 pixel area centered on the home location, and not at home if less than 50% of its bots were in that area. All swarms were given 50 time steps to allow the behavior prescribed by their parameters to emerge; then each task-bot began to run the FLY-HOME algorithm described above. At time step 150, the size of Task-Swarm 2 was reduced to 250 by randomly removing 750 bots, making it half the size of Task-Swarm 1. Thus, at time step 50, the bots in Task-Swarm 1 should fly home, but at time step 150, the bots in Task-Swarm 1 should leave home, while the bots in Task-Swarm 2 should fly home. The success of a comm-swarm was measured by the total time steps of delay (TSD) between the trigger of each arrive-home or leave-home event and the accomplishment of the event. The TSD is defined as: $(t_{\text{AH-1}} - 50) + (t_{\text{LH-1}} - 150) + (t_{\text{AH-2}} - 150)$, where $t_{\text{AH-1}}$ is the time step at which Task-Swarm 1 arrives home, $t_{\text{LH-1}}$ is the time step at which Task-swarm 1 leaves home, and $t_{\text{AH-2}}$ is the time step at which Task-Swarm 2 arrives home.

We designed three swarms for Task-Swarm 1 (a, b, and c) and three swarms for Task-Swarm 2 (a, b, and c), such that each had a qualitatively distinct behavior in terms of the size of the clusters formed by the bots and the dynamics of those clusters. We tested a sample of possible comm-swarms on the nine possible task-swarm pairs. We tested both random comm-swarms, in which there was no explicit separation, cohesion, or alignment specified and the velocity at each step was randomly generated, and non-random comm-swarms that included separation, cohesion, and alignment factors. Our hypothesis was that the more complex dynamics of the non-random swarms (e.g. interacting clusters of bots) could yield more efficient information propagation.

For random comm-swarms, we sampled all possible combinations of three swarm sizes (200, 600, and 1000 bots), three speeds (50, 100, and 200 pixels per time step), and three neighborhood radii (25, 50, and 100 pixels). These values were chosen based on exploratory tests that indicated varied behaviors over these sets of values. These tests also indicated that performance could sometimes be improved if the comm-bots could 1) possibly separate from the task-bots, and 2) possibly cohere and align with the task-bots. Each possible combination of population size, speed, and radius was tried with each of the four possible combinations of these two factors, for a total of 108 possible comm-swarms.

We limited the number of non-random comm-swarms tested by fixing the speed and neighborhood radius values to 200 and 100, respectively, both because these were the optimal values for these two parameters in our tests of random comm-swarms, and because it seemed likely that the largest values for these two parameters would yield better performance. We sampled all possible combinations of three swarm sizes (200, 600, and 1000 bots), three separation strengths (20.0, 60.0, 100.0), three cohesion strengths (0.2, 0.6, 1.0), and three alignment strengths (0.2, 0.6, 1.0). As was the case with random comm-bots, we tested each of these parameter settings with each of four other scenarios: the comm-bots separating (or not) from the task-bots, and 2) cohering and aligning (or not) with the task-bots. This led to a total of 324 possible comm-swarms.

The emergent quality of swarm behavior produces a high variance in observed behavior. This made it difficult to designate any one swarm as the "best" for any given task-swarm pair. For each task-swarm pair, we determined the five random comm-swarms with the lowest average TSDs. There were six comm-swarms that were in this group for five task-swarm pairs; the next best comm-swarms were in that group for only two task-swarm pairs. Furthermore, all six of these better comm-swarms had the same speed (200), neighborhood radius (100), and cohered and aligned with task-bots. The number of comm-bots did not appear to make a difference (we conjecture that an increase in the number of bots results in higher bot counts, which make it more difficult for the relative counts to be reversed when the majorities are reversed), and whether the comm-bots separated from the task-bots was not important (perhaps because the cohesion and alignment with task-bots that are, themselves, separating from each other is sufficient). Further investigation is needed on both issues. We chose to test further the swarm from this group of six that had the lowest TSD in the most swarm pairs (three swarm pairs, compared to one or none for each of the other five). This was the 600 bot comm-swarm that did not separate from task-bots. See Table 1 for results.

No non-random comm-swarm displayed superior performance, even in the weak sense described above for random comm-swarms, but swarms with at least 600 bots appeared to perform somewhat better. For the sake of comparison with the random comm-swarm shown in Table 1, we chose to further investigate non-random comm-swarms of that size and with the same characteristics as that random comm-swarm, but with non-zero separation, cohesion, and alignment strengths. These tests indicated that swarms with higher separation strengths and cohesion and alignment strengths of 0.2 were more effective, leading us to do more extensive testing of the swarm that had separation, cohesion, and alignment strengths of 100.0, 0.2, and 0.2, respectively. See Table 1 for results. The non-random comm-swarm outperforms (in boldface) the random comm-swarm in eight of the nine cases, reducing the TSD by an average of 21.4%. These eight reductions are at a significance level of 0.05 or less (0.0001 in five cases). Observations of a graphical representation of the comm-swarms suggests that the high separation factor between comm-bots serves to amplify the "follow-the-leader" fluctuations induced by the cohesion and alignment between comm-bots, producing a single moving cluster of comm-bots that covers the area quickly and repeatedly, providing better coverage than the more homogeneous coverage of random movement. In both types of comm-swarms, a degree of cohesion and alignment with task-bots appears to be critical; we conjecture that this is necessary to ensure the effective distribution of information.

In the scenario described in Section 1, a comm-swarm would be targeted by the attacking swarm and gradually reduced in size. Thus, it is critical that the comm-swarm be able to function, albeit possibly with reduced effectiveness, with a smaller number of comm-bots. We tested the effectiveness of the two comm-swarms described above with only 60 bots, a 90% reduction in size (see Table 1), and found that the TSDs for these swarms were only, on average, 11.6% larger than their 600-comm-bot counterparts for random comm-swarms, and 18.8% larger than their 600-comm-bot counterparts for non-random comm-swarms, suggesting that these comm-swarms are robust to significant losses.

Table 1. TSD scores for all task-swarm pairs, mean and standard deviation of 20 runs

Swarm Pair	R-CS, 600 bots		NR-CS, 600 bots		R-CS, 60 bots		NR-CS, 60 bots	
	Mean	SD	Mean	SD	Mean	SD	Mean	SD
1a-2a	576.2	52.8	**391.7**	91.3	633.9	67.3	469.3	76.2
1a-2b	445.2	33.0	**322.2**	35.1	595.8	69.7	482.5	75.5
1a-2c	1204.0	147.1	**894.9**	142.3	1162.3	181.6	1011.3	153.0
1b-2a	553.8	98.4	**442.6**	73.3	560.8	81.0	488.5	69.8
1b-2b	384.4	34.2	**274.6**	38.3	493.4	59.4	372.7	58.4
1b-2c	1107.1	97.4	**965.4**	111.9	1127.0	217.1	1132.4	271.9
1c-2a	540.1	71.3	**479.8**	111.7	568.9	87.8	511.4	95.3
1c-2b	409.3	60.1	**355.8**	49.4	493.0	79.2	446.7	97.5
1c-2c	**893.5**	103.5	1040.9	144.1	951.3	112.8	950.3	170.1

R(NR)-CS = Random(Non-Random) Comm-Swarm, SD = Standard Deviation

3 Future Work

Given the preliminary nature of these experiments, there is a great deal of work to be done to refine and explore the capabilities of comm-swarms. For example, preliminary tests suggest that the performance of comm-swarms can be improved by introducing alternating information collection and information distribution phases, each with a separate set of parameters that tune the swarm to that activity. We are investigating this possibility further.

More importantly, we view our current work as the first step in a program to develop a general communication mechanism for cooperating swarms. Given the increasing miniaturization of actual bots, one might design a system of multiple specialized swarms that work together to accomplish a task. One of the challenges in designing such a system would be to provide a mechanism that facilitates information transfer among these swarms. We are currently developing a *communication-link swarm* that will allow information transfer between multiple, mobile task-swarms.

A general measure of the efficiency of information circulation in such a system would be important, allowing us to measure the effectiveness of a comm-swarm in a non-task-specific way. We have begun to develop a measure of the communication efficiency of comm-swarms that is based on the age of the information being distributed, and we are investigating the relationship between this measure and measures developed by others that might be useful in characterizing the effectiveness of comm-swarms: mixing measures [1], measures of information storage and transfer [5], and the moving average Laplacian of [4].

In a different arena, comm-swarms might be useful for particle swarm optimization (PSO), a flocking-inspired optimization technique in which virtual particles search the solution space guided by high-quality solutions found by themselves and by other particles. Communication among particles is critical to the success of the algorithm and we are currently developing a PSO variant that uses comm-swarms to provide an effective communication mechanism.

References

1. Finn, M.D., Cox, S.M., Byrne, H.M.: Mixing measures for two-dimensional chaotic Stokes flow. Journal of Engineering Mathematics 48, 129–155 (2004)
2. Parunak, H.V.D., Purcell, M., O'Connell, R.: Digital pheromones for autonomous coordination of swarming UAVs. American Institute of Aeronautics and Astronautics (2002)
3. Reynolds, C.W.: Flocks, herds and schools: A distributed behavioral model. SIGGRAPH Comput. Graph. 21, 25–34 (1987)
4. Skufca, J.D., Bollt, E.M.: Communication and synchronization in disconnected networks with dynamic topology: Moving neighborhood networks. Mathematical Biosciences and Engineering 1(2), 1–13 (2004)
5. Wang, X.R., Miller, J.M., Lizier, J.T., Prokopenko, M., Rossi, L.F.: Measuring information storage and transfer in swarms. In: Proceedings of the Eleventh European Conference on the Synthesis and Simulation of Living Systems, pp. 838–845. Massachusetts Institute of Technology (2011)

Author Index